Safety and Justice Program

PREDICTIVE POLICING

The Role of Crime Forecasting
in Law Enforcement Operations

Walter L. Perry, Brian McInnis, Carter C. Price,
Susan C. Smith, John S. Hollywood

Supported by the National Institute of Justice

The research described in this report was sponsored by the National Institute of Justice and conducted in the Safety and Justice Program within RAND Justice, Infrastructure, and Environment.

This project was supported by Award No. 2010-IJ-CX-K007, awarded by the National Institute of Justice, Office of Justice Programs, U.S. Department of Justice. The opinions, findings, and conclusions or recommendations expressed in this publication are those of the authors and do not necessarily reflect those of the Department of Justice.

Library of Congress Cataloging-in-Publication Data is available for this publication.

ISBN: 978-0-8330-8148-3

The RAND Corporation is a nonprofit institution that helps improve policy and decisionmaking through research and analysis. RAND's publications do not necessarily reflect the opinions of its research clients and sponsors.

Support RAND—make a tax-deductible charitable contribution at www.rand.org/giving/contribute.html

RAND® is a registered trademark.

RAND OFFICES
SANTA MONICA, CA • WASHINGTON, DC
PITTSBURGH, PA • NEW ORLEANS, LA • JACKSON, MS • BOSTON, MA
DOHA, QA • CAMBRIDGE, UK • BRUSSELS, BE
www.rand.org

Preface

Smart, effective, and proactive policing is clearly preferable to simply reacting to criminal acts. Although there are many methods to help police respond to crime and conduct investigations more effectively, predicting where and when a crime is likely to occur—and who is likely responsible for prior crimes—has recently gained considerable currency. Law enforcement agencies across the United States are employing a range of predictive policing approaches, and much has been written about their effectiveness. This guide for practitioners offers a focused examination of the predictive techniques currently in use, identifies the techniques that show promise if adopted in conjunction with other policing methods, and shares findings and recommendations to inform future research and clarify the policy implications of predictive policing.

This work was sponsored by the National Law Enforcement and Corrections Technology Center of Excellence on Information and Geospatial Technology at the National Institute of Justice (NIJ). This guide will be of interest to law enforcement personnel at all levels and is one in a series of NIJ-sponsored resources for police departments.

The research reported here was conducted in the RAND Safety and Justice Program, which addresses all aspects of public safety and the criminal justice system, including violence, policing, corrections, courts and criminal law, substance abuse, occupational safety, and public integrity. Program research is supported by government agencies, foundations, and the private sector.

This program is part of RAND Justice, Infrastructure, and Environment, a division of the RAND Corporation dedicated to improving policy and decisionmaking in a wide range of policy domains, including civil and criminal justice, infrastructure protection and homeland security, transportation and energy policy, and environmental and natural resource policy.

Questions or comments about this report should be sent to the project leader, John Hollywood (John_Hollywood@rand.org). For more information about the Safety and Justice Program, see http://www.rand.org/safety-justice or contact the director at sj@rand.org.

Contents

Figures

Tables

Summary

Predictive policing is the application of analytical techniques—particularly quantitative techniques—to identify likely targets for police intervention and prevent crime or solve past crimes by making statistical predictions. Several predictive policing methods are currently in use in law enforcement agencies across the United States, and much has been written about their effectiveness. Another term used to describe the use of analytic techniques to identify likely targets is *forecasting*. Although there is a difference between prediction and forecasting, for the purposes of this guide, we use them interchangeably.[1]

Objectives and Approach

Predictive methods allow police to work more proactively with limited resources. The objective of these methods is to develop effective strategies that will prevent crime or make investigation efforts more effective. However, it must be understood at all levels that applying predictive policing methods is not equivalent to finding a crystal ball. For a policing strategy to be considered effective, it must produce tangible results. The objective of this research was to develop a reference guide for departments interested in predictive policing, providing assessments of both the most promising technical tools for making predictions and the most promising tactical approaches for acting on them. More broadly, this guide is intended to put predictive policing in the context of other modern, proactive policing measures.

We approached this task in three ways:

1. We conducted a literature search of academic papers, vendor tool presentations, and recent presentations at conferences, drawing lessons from similar predictive

---

[1] The most common distinction is that forecasting is objective, scientific, and reproducible, whereas prediction is subjective, mostly intuitive, and nonreproducible. According to this distinction, the methods described in this report are essential forecasting methods. However, the law enforcement community has used *predictive policing* to describe these methods, so it is the term favored here.

techniques used in counterinsurgency and counter–improvised explosive devices operations and related research by the U.S. Department of Defense.

2. We reviewed a number of cases of departments using predictive policing techniques that appear promising.

3. We developed a taxonomy of the different types of operational applications that can be supported using predictive policing.

In many cases, we were able to illustrate how predictive technologies are being used to support police operations through a set of examples and case studies. Although some of the methods are promising and describe the current state of field, they are still more academic than practical. Consequently, this guide can also be viewed as a profile of the state of the art of predictive policing practices and the development of new predictive technologies. As such, it can be considered a baseline document.

A Taxonomy of Predictive Methods

In our assessment of predictive policing, we found that predictive methods can be divided into four broad categories:

1. *Methods for predicting crimes:* These are approaches used to forecast places and times with an increased risk of crime.

2. *Methods for predicting offenders:* These approaches identify individuals at risk of offending in the future.

3. *Methods for predicting perpetrators' identities:* These techniques are used to create profiles that accurately match likely offenders with specific past crimes.

4. *Methods for predicting victims of crimes:* Similar to those methods that focus on offenders, crime locations, and times of heightened risk, these approaches are used to identify groups or, in some cases, individuals who are likely to become victims of crime.

Tables S.1–S.4 summarize each category and show the range of approaches that law enforcement agencies have employed to predict crimes, offenders, perpetrators' identities, and victims, respectively. We found a near one-to-one correspondence between conventional crime analysis and investigative methods and the more recent "predictive analytics" methods that mathematically extend or automate the earlier methods. Conventional methods tend to be heuristic, or mathematically simple. As a result, they are low-cost and can work quite well, especially for analysts facing *low to moderate* data volumes and levels of complexity. In contrast, full-scale predictive analytics require sophisticated analysis methods that draw on *large* data sets. In this context, *large* refers to an amount of data beyond what a single analyst could recall without the assistance of a computer program or similar resources. Conversely, *low to moderate* refers to a data set that is sufficiently small that an analyst could reasonably recall its key facts.

Table S.1 summarizes predictive policing methods related to predicting crimes. As the table shows, conventional approaches start with mapping crime locations and determining (using human judgment) where crimes are concentrated ("hot spots"). The corresponding predictive analytics methods start, at the most basic level, with regression analyses and extend all the way to cutting-edge mathematical models that are the subjects of active research.

Table S.2 summarizes methods to identify individuals at high risk of offending in the future. The bulk of these methods relate to assessing individuals' risk. Here, conventional methods rely on clinical techniques that add up the number of risk factors to create an overall risk score. The corresponding predictive analytics methods use regression and classification models to associate the presence of risk factors with a percent chance that a person will offend. Also of interest are methods that identify criminal groups (especially gangs) that are likely to carry out violent assaults on each other in the near future. Hence, these methods can also be used to assess the risk that an individual will become a victim of crime.

Table S.3 summarizes methods used to identify likely perpetrators of past crimes. These approaches are essentially real-world versions of the board game Clue™: They use available information from crime scenes to link suspects to crimes, both directly and by processes of elimination. In conventional approaches, investigators and analysts

Table S.1
Law Enforcement Use of Predictive Technologies: Predicting Crimes

Problem	Conventional Crime Analysis (low to moderate data demand and complexity)	Predictive Analytics (large data demand and high complexity)
Identify areas at increased risk		
Using historical crime data	Crime mapping (hot spot identification)	Advanced hot spot identification models; risk terrain analysis
Using a range of additional data (e.g., 911 call records, economics)	Basic regression models created in a spreadsheet program	Regression, classification, and clustering models
Accounting for increased risk from a recent crime	Assumption of increased risk in areas immediately surrounding a recent crime	Near-repeat modeling
Determine when areas will be most at risk of crime	Graphing/mapping the frequency of crimes in a given area by time/date (or specific events)	Spatiotemporal analysis methods
Identify geographic features that increase the risk of crime	Finding locations with the greatest frequency of crime incidents and drawing inferences	Risk terrain analysis

Table S.2
Law Enforcement Use of Predictive Technologies: Predicting Offenders

Problem	Conventional Crime Analysis (low to moderate data demand and complexity)	Predictive Analytics (large data demand and high complexity)
Find a high risk of a violent outbreak between criminal groups	Manual review of incoming gang/ criminal intelligence reports	Near-repeat modeling (on recent intergroup violence)
Identify individuals who may become offenders:	Clinical instruments that summarize known risk factors	Regression and classification models using the risk factors
Probationers and parolees at greatest risk of reoffending		
Domestic violence cases with a high risk of injury or death		
Mental health patients at greatest risk of future criminal behavior or violence		

Table S.3
Law Enforcement Use of Predictive Technologies: Predicting Perpetrator Identities

Problem	Conventional Crime Analysis (low to moderate data demand and complexity)	Predictive Analytics (large data demand and high complexity)
Identify suspects using a victim's criminal history or other partial data (e.g., plate number)	Manually reviewing criminal intelligence reports and drawing inferences	Computer-assisted queries and analysis of intelligence and other databases
Determine which crimes are part of a series (i.e., most likely committed by the same perpetrator)	Crime linking (use a table to compare the attributes of crimes known to be in a series with other crimes)	Statistical modeling to perform crime linking
Find a perpetrator's most likely anchor point	Locating areas both near and between crimes in a series	Geographic profiling tools (to statistically infer most likely points)
Find suspects using sensor information around a crime scene (GPS tracking, license plate reader)	Manual requests and review of sensor data	Computer-assisted queries and analysis of sensor databases

do this largely by tracing these links manually, with assistance from simple database queries (usually, the names, criminal records, and other information known about the suspects). Predictive analytics automate the linking, matching available "clues" to potential (and not previously identified) suspects across very large data sets.

Table S.4 summarizes methods to identify groups—and, in some cases, individuals—who are likely to become victims of crime. These methods mirror those used to predict where and when crimes will occur, as well as some of the methods used to predict who is most likely to commit crimes. Predicting victims of crime requires

Table S.4
Law Enforcement Use of Predictive Technologies: Predicting Crime Victims

Problem	Conventional Crime Analysis (low to moderate data demand and complexity)	Predictive Analytics (large data demand and high complexity)
Identify groups likely to be victims of various types of crime (vulnerable populations)	Crime mapping (identifying crime type hot spots)	Advanced models to identify crime types by hot spot; risk terrain analysis
Identify people directly affected by at-risk locations	Manually graphing or mapping most frequent crime sites and identifying people most likely to be at these locations	Advanced crime-mapping tools to generate crime locations and identify workers, residents, and others who frequent these locations
Identify people at risk for victimization (e.g., people engaged in high-risk criminal behavior)	Review of criminal records of individuals known to be engaged in repeated criminal activity	Advanced data mining techniques used on local and other accessible crime databases to identify repeat offenders at risk
Identify people at risk of domestic violence	Manual review of domestic disturbance incidents; people involved in such incidents are, by definition, at risk	Computer-assisted database queries of multiple databases to identify domestic and other disturbances involving local residents when in other jurisdictions

identifying at-risk groups and individuals—for example, groups associated with various types of crime, individuals in proximity to at-risk locations, individuals at risk of victimization, and individuals at risk of domestic violence.

Prediction-Led Policing Process and Prevention Methods

Making "predictions" is only half of prediction-led policing; the other half is carrying out interventions, acting on the predictions that lead to reduced crime (or at least solve crimes). What we have found in this study is that predictive policing is best thought of as part of a *comprehensive business process*. That process is summarized in Figure S.1. We also identified some emerging practices for running this business process successfully through a series of discussions with leading predictive policing practitioners.

At the core of the process shown in Figure S.1 is a four-step cycle (top of figure). The first two steps are collecting and analyzing crime, incident, and offender data to produce predictions. Data from disparate sources in the community require some form of data fusion. Efforts to combine these data are often far from easy, however.

The third step is conducting police operations that intervene against the predicted crime (or help solve past crimes). The type of intervention will vary with the situation and the department charged with intervening. Figure S.1 shows three broad types of interventions (lower right of figure). They are, from simplest to most complicated,

Figure S.1
The Prediction-Led Policing Business Process

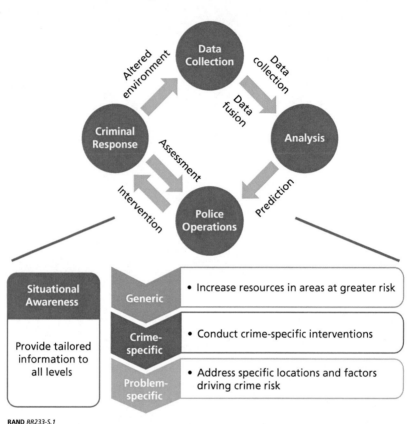

RAND *RR233-S.1*

generic intervention, crime-specific intervention, and problem-specific intervention. In general, we hypothesize that the more complicated interventions will require more resources, but they will be better tailored to the actual crime problems—and get better results. Regardless of the type of intervention, those carrying it out need information to execute the intervention successfully. Thus, providing information that fills the need for *situational awareness* among officers and staff is a critical part of any intervention plan.

The interventions lead to a criminal response that ideally reduces or solves crime (the fourth step). In the short term, an agency needs to do rapid *assessments* to ensure that the interventions are being implemented properly and that there are no immediately visible problems. The longer-term criminal response is measured through changes in the collected data, which, in turn, drives additional analysis and modified operations, and the cycle repeats.

Predictive Policing Myths and Pitfalls

Many types of analysis that inform predictive policing have been widely used in law enforcement and other fields, just under different names. The lessons from these prior applications can highlight many well-known pitfalls that can lead practitioners astray and can provide recommendations to enhance the effectiveness of predictive policing efforts.

Predictive Policing Myths

"Predictive policing" has received a substantial amount of attention in the media and the research literature. However, some myths about these techniques have also propagated. This is partly a problem of unrealistic expectations: Predictive policing has been so hyped that the reality cannot live up to the hyperbole. There is an underlying, erroneous assumption that advanced mathematical and computational power is both necessary and sufficient to reduce crime. Here, we dispel four of the most common myths about predictive policing:

- *Myth 1: The computer actually knows the future.* Some descriptions of predictive policing make it sound as if the computer can foretell the future. Although much news coverage promotes the meme that predictive policing is a crystal ball, these algorithms predict the risk of future events, not the events themselves. The computer, as a tool, can dramatically simplify the search for patterns, but all these techniques are extrapolations from the past in one way or another. In addition, predictions are only as good as the underlying data used to make them.
- *Myth 2: The computer will do everything for you.* Although it is common to promote software packages as end-to-end solutions for predictive policing, humans remain—by far—the most important elements in the predictive policing process. Even with the most complete software suites, humans must find and collect relevant data, preprocess the data so they are suitable for analysis, design and conduct analyses in response to ever-changing crime conditions, review and interpret the results of these analyses and exclude erroneous findings, analyze the integrated findings and make recommendations about how to act on them, and take action to exploit the findings and assess the impact of those actions.
- *Myth 3: You need a high-powered (and expensive) model.* Most police departments do not need the most expensive software packages or computers to launch a predictive policing program. Functionalities built into standard workplace software (e.g., Microsoft Office) and geographic information systems (e.g., ArcGIS) can support many predictive methods. Although there is usually a correlation between the complexity of a model and its predictive power, increases in predictive power have tended to show diminishing returns. Simple heuristics have been found to be nearly as good as analytic software in performing some tasks. This finding is

especially important for small departments, which often have insufficient data to support large, sophisticated models.

- *Myth 4: Accurate predictions automatically lead to major crime reductions.* Predictive policing analysis is frequently marketed as the path to the end of crime. The focus on the analyses and software can obscure the fact that predictions, on their own, are just that—predictions. Actual decreases in crime require taking action based on those predictions. Thus, we emphasize again that predictive policing is not about making predictions but about the end-to-end process.

Predictive Policing Pitfalls

To be of use to law enforcement, predictive policing methods must be applied as part of a comprehensive crime prevention strategy. And to ensure that predictive methods make a significant contribution, certain pitfalls need to be avoided:

- *Pitfall 1: Focusing on prediction accuracy instead of tactical utility.* Suppose an analyst is asked to provide predictions of robberies that are as "accurate" as possible (i.e., to design an analysis in which as many future crimes as possible fall inside areas predicted to be high-risk, thus confirming that these areas are high-risk). One way to accomplish this is to designate the entire city a giant "risk area." However, such a designation has almost no tactical utility. Identifying a hot spot that is the size of a city may be accurate, but it does not provide any information that police officers do not already have. To ensure that predicted hot spots are small enough to be actionable, we must accept some limits on "accuracy" as measured by the proportion of crimes occurring in the hot spots.

- *Pitfall 2: Relying on poor-quality data.* There are three typical deficiencies that can affect data quality: data censoring, systematic bias, and relevance. Data censoring involves omitting data for incidents of interest in particular places (and at particular times). If the data are censored, it will appear that there is no crime in a given areas. Systematic bias can result from how the data are collected. For example, if especially heavy burglary activity is reported between 7:00 and 8:00 a.m., it may not be immediately clear whether a large number of burglaries actually occurred during that hour or whether that was when property owners and managers discovered and reported burglaries that took place overnight. Finally, relevance refers to the usefulness of the data. For some crime clusters, it can be very useful to have data going back many months or years. Conversely, if there is a spree of very similar robberies likely committed by the same criminal, several months of data will not be of much use.

- *Pitfall 3: Misunderstanding the factors behind the prediction.* Observers—especially practitioners tasked with making hot spots go away—may reasonably ask, "For a given hot spot, what *factors* are driving risk?" "The computer said so" is far from an adequate answer. In general, predictive tools are designed in a way that makes it difficult, if not impossible, to highlight the risk factors present in specific areas. There has been some improvement, however. When applying techniques, such as regression or any of the data mining variants, using common sense to vet the factors incorporated into the model will help avoid spurious relationships.

- *Pitfall 4: Underemphasizing assessment and evaluation.* During our interviews with practitioners, very few said that they had evaluated the effectiveness of the predictions they produced or the interventions developed in response to their predictions. The effectiveness of any analysis and interventions should be assessed as part of the overall effort to keep the data current. Measurements are key to identifying areas for improvement, modifying interventions, and distributing resources.
- *Pitfall 5: Overlooking civil and privacy rights.* The very act of labeling areas and people as worthy of further law enforcement attention inherently raises concerns about civil liberties and privacy rights . Labeling areas as "at-risk" appears to pose fewer problems because, in that case, individuals are not being directly targeted. The U.S. Supreme Court has ruled that standards for what constitutes reasonable suspicion are relaxed in "high-crime areas" (e.g., hot spots). However, what formally constitutes a "high-crime" area, and what measures may be taken in such areas under "relaxed" reasonable-suspicion rules, is an open question.

Recommendations

Our conclusions center on advice to three communities: police departments (the buyer), vendors and developers, and crime fighters. Our advice centers on the role of predictive policing in the larger context of law enforcement operations.

Advice for Buyers (Law Enforcement Agencies)

All departments can benefit from predictive policing methods and tools; the distinction is in how sophisticated (and expensive) these tools need to be. In thinking about these needs, it is important to remember that the value of predictive policing tools is in their ability to provide *situational awareness* of crime risks and the information needed to act on those risks and preempt crime. The question, then, is which set of tools can best provide the situational awareness a department needs?

Small agencies with relatively few crimes per year and with reasonably understandable distributions of crime (e.g., a jurisdiction with a few shopping areas that are the persistent hot spots) are unlikely to need much more than core statistical and display capabilities. These tools are available for free or at low cost and include built-in capabilities in Microsoft Office, basic geographic information tools, base statistics packages, and perhaps some advanced visualization tools, such as the National Institute of Justice–sponsored CrimeStat series.

Larger agencies with large volumes of incident and intelligence data that need to be analyzed and shared will want to consider more sophisticated and, therefore, more costly systems. It is helpful to think of these as enterprise information technology systems that make sense of large data sets to provide situational awareness across a department (extending, in many cases, to the public). These systems should help agencies understand the where, when, and who of crime and identify the specific problems that drive crime in order to take action against them. Key considerations include interoperability with the department's records management, computer-aided dispatch, and other systems; the ability to incorporate "intelligence" tips from officers (e.g., via field interviews) and the public; the types of displays ("dashboards") and supporting information the system can provide; and, of course, the types of analyses and predictions the system can support and under what conditions.

Advice for Vendors and Developers

The list of questions for purchasers doubles as guidance on desired capabilities for those who develop predictive tools. Looking ahead, it could be useful to move beyond predictions to offer explicit decision support for resource allocation and other decisions.

We emphasize that predictive policing tools and methods are very useful, but they are not crystal balls. Media reports and advertisements can give an impression that one merely needs to ask a computer where and when to go to catch criminals in the act. We ask that vendors be accurate in describing their systems as identifying crime risks, not foretelling the future.

Finally, developers must be aware of the major financial limitations that law enforcement agencies face in procuring and maintaining new systems. Licensing fees of into the millions of dollars are simply not affordable for most departments. We suggest that vendors consider business models that can make predictive policing systems more affordable for smaller agencies, such as regional cost sharing.

Advice for Crime Fighters

Generating predictions is just half of the predictive policing business process; taking actions to interdict crimes is the other half. The specific interventions will vary by objective and situation. (A number of examples are described in Chapters Three and Four of this guide; core resources on interventions are the Office of Justice Programs' CrimeSolutions.gov and the Center on Problem Oriented-Policing.) However, we have identified some promising features of successful intervention efforts:

- There is substantial top-level support for the effort.
- Resources are dedicated to the task.
- The personnel involved are interested and enthusiastic.
- Efforts are made to ensure good working relationships between analysts and officers.

- The predictive policing systems and other department resources provide the shared situational awareness needed to make decisions about where and how to take action.
- Synchronized support is provided when needed.
- Responsible officers have the freedom to carry out interventions and accountability for solving crime problems.
- The interventions are based on building good relationships with the community and good information ("intelligence").

Designing intervention programs that take these attributes into account, in combination with solid predictive analytics, can go a long way toward ensuring that predicted crime risks do not become real crimes.

Acknowledgments

We are grateful for the assistance provided by the professional staff of several police departments in the United States and Canada with whom we engaged in comprehensive discussions of the predictive and information-led policing approaches employed in their departments: the Metropolitan Nashville Police Department in Nashville, Tennessee; the Vancouver Police Department in Vancouver, British Columbia; the Glendale Police Department in Glendale, Arizona; the East Orange Police Department in East Orange, New Jersey; the Minneapolis Police Department in Minneapolis, Minnesota; the Charlotte-Mecklenburg Police Department in Charlotte and Mecklenburg County, North Carolina; and the Baltimore Police Department in Baltimore, Maryland.

We also thank Craig Uchida, president of Justice & Security Strategies, Inc., for his insights on the current practice of predictive policing and for his thoughtful review of this document. In addition, we wish to acknowledge our RAND colleague, Joel Predd, whose helpful comments on our original draft greatly improved this document. Finally, we thank Steve Schuetz, William Ford, and Joel Hunt at the National Institute of Justice for their helpful insights and strong support of this project.

Abbreviations

AHS	Actionable HotSpot
ARIMA	autoregressive integrated moving average
ASI	Addiction Severity Index
BDI-II	Beck Depression Inventory–II
CAD	computer-aided dispatch
CNEP	Cincinnati Neighborhood Enhancement Program
CODEFOR	computer-optimized deployment–focus on results
COINCOP	Counterinsurgency Common Operational Picture
CompStat	comparative statistics
COP	common operational picture
CRUSH	Crime Reduction Utilizing Statistical History
DDACTS	data-driven approaches to crime and traffic safety
DUI	driving under the influence
DWI	driving while intoxicated
FBI	Federal Bureau of Investigation
FY	fiscal year
GIS	geographic information systems
GLM	generalized linear model
GPS	Global Positioning System
IACA	International Association of Crime Analysts

IED	improvised explosive device
IT	information technology
JJIS	Juvenile Justice Information System
KDE	kernel density estimation
LAPD	Los Angeles Police Department
LEIU	Association of Law Enforcement Intelligence Units
LISA	local indicators of spatial association
LS/CMI	Level of Service/Case Management Inventory
LSI-R	Level of Service Inventory–Revised
LST-GAM	local spatiotemporal generalized additive model
MAST	Michigan Alcoholism Screening Test
MMPI-2	Minnesota Multiphasic Personality Inventory–2
MO	modus operandi
NCISP	National Criminal Intelligence Sharing Plan
N-DEx	National Data Exchange
NIJ	National Institute of Justice
Nnh	nearest neighbor hierarchical clustering
NPS	Neighborhood Problem Solving
OST	Office of Science and Technology, National Institute of Justice
PAI	prediction accuracy index
PCL-R	Hare Psychopathy Checklist–Revised
PILOT	Predictive Intelligence–Led Operational Targeting
RMS	records management system
R-PACT	Residential Positive Achievement Change Tool
RRI	recapture rate index
RTM	risk terrain modeling
SARA	Scanning, Analysis, Response, and Assessment (model)

SASSI	Substance Abuse Subtle Screening Inventory
SEATS	Signal Extraction in ARIMA Time Series
SOM	self-organizing map
SPCS	State Plane Coordinate System
SPSS	Statistical Package for Social Sciences
ST-GAM	spatiotemporal generalized additive model
TRAMO	Time-Series Regression with ARIMA Noise, Missing Observations, and Outliers
UCR	Uniform Crime Report

CHAPTER ONE

Introduction

The race is not always to the swift nor the battle to the strong, but that's the way to bet.

—Damon Runyon

Smart, effective, and proactive policing is clearly preferable to simply reacting to criminal acts. Although there are many methods aimed at preventing crime, predicting where and when a crime is likely to occur, who is likely responsible for prior crimes, and who is most likely to offend or be victimized in the future has recently gained considerable currency. Law enforcement agencies across the United States are employing a range of predictive policing methods, and much has been written about their effectiveness. This guide offers a focused examination of predictive techniques currently in use, identifies the techniques that show promise if adopted in conjunction with other policing methods, and shares findings and recommendations to inform future research and clarify the policy implications of predictive policing.

Another term associated with anticipating future criminal acts is *forecasting*. We use this term in the title of this guide alongside *prediction*. There is, in fact a difference between the two. Forecasting is considered objective, scientific, reproducible, and free from individual bias and error. Prediction is thought of as subjective, mostly intuitive, nonreproducible, and subject to individual bias. Because it is scientific, error analysis is possible with forecasts but not with predictions. In line with this distinction, this guide is about forecasting and not prediction. However, the law enforcement community uses the term *predictive policing* and not *forecast policing*. For this reason, for the purposes of this guide, we make no distinction between the two terms.[1]

What Is Predictive Policing?

Predictive policing is the application of analytical techniques—particularly quantitative techniques—to identify likely targets for police intervention and prevent crime or

---

[1] For additional distinctions between forecasting and prediction, see Transtutors, "Difference Between Forecasting and Prediction," undated.

solve past crimes by making statistical predictions. The use of statistical and geospatial analyses to forecast crime levels has been around for decades. In recent years, however, there has been a surge of interest in analytical tools that draw on very large data sets to make predictions in support of crime prevention. These tools greatly increase police departments' reliance on information technology (IT) to collect, maintain, and analyze those data sets, however.

These analytical tools, and the IT that supports them, are largely developed by and for the commercial world. Universities and technology companies have created computer programs based on private-sector models of forecasting consumer behavior. Businesses use predictive analytics to determine sales strategies. For example, Walmart analyzes weather patterns to determine what it stocks in stores, overstocking duct tape, bottled water, and strawberry Pop-Tarts before major weather events. The first two items are expected, but the Pop-Tarts represent a "non-obvious relationship."[2] These relationships are uncovered through statistical analyses of previous customer purchases during similar major weather events.

Many similar relationships in law enforcement can be explored with predictive policing. Police agencies use computer analysis of information about past crimes, the local environment, and other pertinent intelligence to "predict" and prevent crime. The idea is to improve situational awareness at the tactical and strategic levels and to develop strategies that foster more efficient and effective policing. With situational awareness and anticipation of human behavior, police can identify and develop strategies to prevent criminal activity by repeat offenders against repeat victims. These methods also allow police departments to work more proactively with limited resources.

However, it must be understood at all levels that applying these methods is not equivalent to finding a crystal ball. For a policing strategy to be considered effective, it must produce tangible results. For example, crime rates should be lower, arrest rates for serious offenses should increase, and there should be an observable positive impact on social and justice outcomes.

A Criminological Justification for Predictive Policing: Why Crime Is "Predictable"

There is a strong body of evidence to support the theory that crime is predictable (in the statistical sense)—mainly because criminals tend to operate in their comfort zone. That is, they tend to commit the type of crimes that they have committed successfully in the past, generally close to the same time and location. Although this is not universally true, it occurs with sufficient frequency to make these methods work reasonably

---

[2] Olivia Katrandjian, "Hurricane Irene: Pop-Tarts Top List of Hurricane Purchases," ABC News, August 27, 2011.

well. According to Jeff Brantingham, an anthropologist who helps supervise the predictive policing project for the Los Angeles Police Department (LAPD) at the University of California, Los Angeles,

> The naysayers want you to believe that humans are too complex and too random— that this sort of math can't be done . . . but humans are not nearly as random as we think. . . . In a sense, crime is just a physical process, and if you can explain how offenders move and how they mix with their victims, you can understand an incredible amount.[3]

Brantingham's remarks are supported by major theories of criminal behavior, such as routine activity theory, rational choice theory, and crime pattern theory.[4] For this study, we consolidated these theories into what we refer to as a *blended theory*:

- Criminals and victims follow common life patterns; overlaps in those patterns indicate an increased likelihood of crime.
- Geographic and temporal features influence the where and when of those patterns.
- As they move within those patterns, criminals make "rational" decisions about whether to commit crimes, taking into account such factors as the area, the target's suitability, and the risk of getting caught.

The theoretical justification for predictive policing, then, is that we can identify many of these patterns and factors through analytics and then can steer criminals' decisions to prevent crimes with tactical interventions.

The blended theory best fits "stranger offenses," such as robberies, burglaries, and thefts. It is less applicable to vice and relationship violence, both of which involve human connections that both extend beyond limited geographic boundaries and lead to decisions that do not fit into traditional "criminal rational choice" frameworks. Nonetheless, alternative theories have been tested to explain vice and relationship violence, leading to the development of instruments and methods for assessing risks in these areas as well.

A Brief History of Predictive Policing

Although methods aimed at predicting crime have been around for a long time, it is only recently that modern technology has moved these attempts from simple heuristic

---

[3] Quoted in Joel Rubin, "Stopping Crime Before It Starts," *Los Angeles Times*, August 21, 2010.

[4] For summaries of criminal behavior theories, see Ronald V. Clarke and Marcus Felson, eds., *Routine Activity and Rational Choice*, New Brunswick, N.J.: Transaction Publishers, 2003.

methods to sophisticated mathematical algorithms. In this section, we trace the modern history of predictive policing and the evolution of a training regimen for practitioners.

Background

Police Chief (ret.) William J. Bratton and the LAPD are credited with envisioning the predictive policing model. By 2008, Chief Bratton had spoken widely in the public arena about the successes of the LAPD, including the department's recent introduction of predictive analytics to anticipate gang violence and to support real-time crime monitoring. Chief Bratton suggested that this new approach could build on and enhance existing approaches, including community-oriented policing and intelligence-led policing.[5]

In 2008, Chief Bratton worked closely with the acting director of the Bureau of Justice Assistance, James H. Burch II, and the acting director of the National Institute of Justice (NIJ), Kristina Rose, to explore the new concept of predictive policing and its implications for law enforcement agencies. This effort involved gathering expert researchers, practitioners, government officials, and law enforcement leaders at two consecutive predictive policing symposiums hosted by NIJ.[6]

In her opening remarks at the first symposium, held in Los Angeles in November 2009, Kristina Rose acknowledged Chief Bratton as having "served as the catalyst for bringing predictive policing to the forefront."[7] She also acknowledged industry-wide interest in understanding the term *predictive policing*, as well as the policy, technical, and operational implications of predictive policing approaches. She named several agencies across the United States, including the Boston, Chicago, Los Angeles, D.C. Metropolitan, New York, and Shreveport police departments and the Maryland State Police who had responded to a solicitation for proposals from agencies interested in taking part in a predictive policing demonstration initiative.

The Los Angeles symposium was widely discussed within law enforcement circles at all levels, and the topic of predictive policing generated a great deal of interest across traditional and social media. Consultants and private companies quickly began providing professional services and software that they deemed useful and appropriate for predictive policing efforts.

With momentum strong, a second symposium was held in June 2010 in Providence, Rhode Island. The event featured extended discussions from the first symposium, with a general consensus that continued exploration of predictive policing was needed. Challenges, successes, limitations, and scalability were major points of focus.

---

5   "What Is Predictive Policing?" *WP's Police Tech*, March 16, 2012.

6   "Transcript: Perspectives in Law Enforcement—The Concept of Predictive Policing: An Interview with Chief William Bratton," U.S. Department of Justice, Bureau of Justice Assistance, November 2009.

7   Kristina Rose, Acting Director, National Institute of Justice, "Predictive Policing Symposium: Opening Remarks," transcript of speech at Predictive Policing Symposium, Los Angeles, Calif., November 19, 2009.

Participants emphasized the critical need for data sharing and regionalization, as well as the need for a strong analytical capability.[8]

The next two years saw an explosion of interest in predictive policing. Much of this interest came from heavy media coverage of what is now called PredPol, a prediction software package used in Santa Cruz and Los Angeles, California. *CBS Evening News,* the *New York Times,* and *NBC Nightly News,* among others, have covered PredPol specifically and predictive policing in general.[9] Much of the coverage implied (or directly stated) that the software could literally predict where crime would occur and used as an example the arrest of two suspects in a parking garage who had been designated high-risk. A commercial for IBM's predictive analytics showing a police officer deployed (through data analysis) to a convenience store just before a would-be robber arrives also attracted significant attention.[10]

By the time of this writing, the catch phrase "predictive policing" had caught on in law enforcement. Researchers, educators, government officials, consultants, crime analysts, police commanders, private software companies, civil rights activists, and the media have all made their interests and positions on the topic well known, and it continues to be hotly debated both inside and outside policing circles.

There are a number of stakeholders worth noting in the field of predictive policing. Researchers have the background and expertise to design predictive models; civil rights activists are concerned that these techniques may intrude on the rights of citizens, especially poor and minority populations; analysts and investigators have a professional interest in how these approaches can improve their work and make it more useful; police chiefs are eager to find new techniques to reduce crime without adding to their workforce; and the private sector sees potential funding from research grants, consulting, and software development. With the onset of irregular warfare, the U.S. military has developed an interest in predictive techniques as well. Insurgent groups behave much like organized criminal gangs, so policing methods, including predictive analyses, could be useful for military applications. Companies and federal agencies responsible for site security are also interested in predictive analyses to help them use their resources more efficiently and effectively to defend against the likeliest threats.

Training

Over much of the past 40 years, crime analysts have struggled to get formal training in "hot spot" identification. NIJ recognized the need to train analysts, researchers, and

---

[8] National Institute of Justice, "Predictive Policing Symposiums: What Chiefs Expect from Predictive Policing—Perspectives from Police Chiefs," web page, last updated January 6, 2012.

[9] "'Predictive Policing' Making LA Safer," video, *CBS Evening News,* April 11, 2012; Erica Goode, "Sending the Police Before There's a Crime," *New York Times,* August 15, 2011; "New Police Motto: To Predict and Serve?" video, *NBC Nightly News,* August 20, 2011.

[10] IBM, "Predictive Analytics—Police Use Analytics to Reduce Crime," video advertisement, March 27, 2012.

students in spatial analysis and crime mapping, and it developed the Crime Mapping Research Center to respond to that need. Later renamed the Mapping and Analysis for Public Safety program, it serves as a vehicle for the NIJ to provide research, training, and periodic conferences to inform and train those interested in this emerging area.

The mid-1990s saw the establishment of the Crime Mapping and Analysis Program, an outgrowth of the National Law Enforcement and Corrections Technology Center–Rocky Mountain, to provide hands-on training classes to law enforcement analysts in the use of both ESRI's ArcGIS and MapInfo software, the field's two most popular software programs. Later, between 2008 and 2010, NIJ again invested in training for analysts by commissioning a CrimeStat III workbook and training class, which was offered both in person and online. CrimeStat III generates a number of crime analysis algorithms that analysts can plug into ArcGIS and other geographic information system (GIS) packages, the more sophisticated of which can be considered predictive policing tools.

NIJ decided to end its direct training in 2011. At that time, the International Association of Crime Analysts (IACA) took over the training of analysts and continues to provide these classes both online and in person. The association published a white paper in July 2012 on GIS standards in law enforcement, which addressed the requirements of mapping software. It focused, in particular, on the current state of GIS in crime analysis, the functionality required for its use in crime analysis, and relevant literature on related topics and key readings on spatial analysis, crime mapping, and hot spot analysis. In the future, it is likely that the IACA and other practitioner associations will develop courses and standards in other areas of predictive policing, depending on practitioner demand, possibly with federal support. Furthermore, this document is intended as a guide to predictive policing for practitioners and policymakers and will likely also contribute to future training.

Study Objectives and Methods

Understandably, there has been a surge of interest in predictive policing in the media, from practitioners, and from vendors. Even a slight possibility that one method or another can help identify where and when crimes will be committed has led to a great deal of hype. For example, the IACA (of which one of this guide's authors is president) routinely receives email inquiries from practitioners asking where to get tools that will predict crimes and how to implement them in their agencies. Clearly, the field stands to benefit from an effective assessment of these technologies that will allow departments to evaluate how well a given tool can meet their needs.

Objectives

The objective of this study was to develop a reference guide for departments interested in predictive policing, providing assessments of both the most promising technical tools for making predictions and the most promising tactical approaches to act on those predictions. More broadly, this guide is intended to put predictive policing in the context of other modern, proactive policing measures: Predictive policing can be a very useful tool, but it is just that—a tool. It is not a crystal ball.

This guide is one in a series of NIJ-supported resources for police departments. Specifically, it builds on earlier NIJ publications on crime mapping and analysis, including *Mapping Crime: Understanding Hot Spots* and *Mapping Crime: Principle and Practice.*[11]

Throughout this guide, we illustrate with examples and case studies how predictive technologies are being used to support police operations. Although some of the methods are promising and reflect the current state of play, they are still more academic than practical. Consequently, this guide can also be viewed as a description of the state of the art in the practice of predictive policing and in the development of new predictive technologies. As such, it can be considered a baseline document.

Approach

We approached the task of developing this guide by conducting an extensive literature review, reviewing several case studies, and developing a taxonomy of operational applications of the methods identified over the course of our study.

- *Extended literature review:* We examined not only academic papers but also vendor tool presentations and very recent presentations at conferences. We also looked at what the U.S. Department of Defense has done in the areas of counterinsurgency and counter–improvised explosive device (IED) operations, an effort that included an examination of recent RAND research on these topics.
- *Case studies:* We reviewed cases of departments using predictive policing techniques that appear promising. We emphasize "promising" because, with a few exceptions, there have been no formal controlled evaluations (though some were under way at the time of this writing). We reviewed papers and presentations by department personnel and engaged in discussions with them. We then qualitatively analyzed the discussion and presentation notes to identify common themes regarding promising practices.
- *Taxonomy development:* From the previous two steps, we developed a taxonomy of operational applications that can be supported by predictive policing, as well as the specific methods and interventions that support those applications. We

---

[11] John E. Eck, Spencer Chainey, James G. Cameron, Michael Leitner, and Ronald E. Wilson, *Mapping Crime: Understanding Hot Spots*, Washington, D.C.: National Institute of Justice, August 2005; Keith Harries, *Mapping Crime: Principle and Practice*, Washington, D.C.: National Institute of Justice, 1999.

identified promising practices (technical and operational) for each "node" in the taxonomy. We also identified some overall pitfalls (and mitigations) and policy issues that need to be addressed.

The Nature of Predictive Policing: This Is Not *Minority Report*

There is an obvious appeal to being able to prevent crime as opposed to merely apprehending offenders after a crime has been committed. For law enforcement agencies, the ability to predict a crime and stop it before it is committed is tantalizing indeed—as it is to the general public. Any hype must be tempered somewhat by considerations of privacy and civil rights, however. Predictive methods, themselves, may not expose sufficient probable cause to apprehend a suspected offender.[12] "Predictions" are generated through statistical calculations that produce estimates, at best; like all techniques that extrapolate the future based on the past, they assume that the past is prologue. Consequently, the results are probabilistic, not certain. Earlier in this chapter, we noted the heavy media coverage of PredPol and the preemptive arrests of car burglary suspects in a garage. In reality, the software directed the arresting officers to the area because it had been the location of recent car burglaries.

Chapters Two and Three explore in greater detail the major statistical methods used by law enforcement, what they do, and where they might be most applicable. We also discuss both academic studies and the experiences of practitioners who have used these approaches.

A Taxonomy of Predictive Methods

In this section, we introduce our taxonomy of predictive policing methods, which organizes the discussion in Chapters Two and Three. In our assessment of predictive policing, we found that predictive methods can be divided into four broad categories:

1. *Methods for predicting crimes:* These are approaches used to forecast places and times with an increased risk of crime.
2. *Methods for predicting offenders:* These approaches identify individuals at risk of offending in the future.
3. *Methods for predicting perpetrators' identities:* These techniques are used to create profiles that accurately match likely offenders with specific past crimes.
4. *Methods for predicting victims of crimes:* Similar to those methods that focus on offenders, crime locations, and times of heightened risk, these approaches are used

---

[12] We take up the issues of privacy and civil rights in Chapter Three.

to identify groups or, in some cases, individuals who are likely to become victims of crime.

Chapter Two discusses methods associated with predicting potential crimes, and Chapter Three focuses on methods associated with predicting offenders.

Tables 1.1–1.4 summarize each category and show the range of approaches that law enforcement agencies have employed to predict crimes, offenders, perpetrators' identities, and victims, respectively. In assessing predictive methods, we found a near one-to-one correspondence between conventional crime analysis and investigative methods and the more recent "predictive analytics" methods that mathematically extend or automate the earlier methods. The conventional methods tend to be manual, heuristic, or mathematically simple, rarely relying on a tool more complicated than a spreadsheet or a basic GIS program. As a result, they are low-cost, but they can work quite well, especially for analysts facing *low to moderate* data volumes and levels of complexity. In contrast, full-scale predictive analytics require sophisticated analysis systems that draw on *large* data sets.[13] The analytics and supporting database systems tend to be high-cost (though there are open-source options, too) and thus tend to be reasonably well matched to large agencies with large and complex volumes of crime, incident, and offender data.

Table 1.1 summarizes predictive policing methods related to predicting crimes, i.e., identifying places and times that correspond to an increased risk of crime. As the table shows, conventional approaches start with mapping crime locations and determining (using human judgment) where crimes are concentrated (the hot spots). These approaches might include making bar graphs showing when crimes have occurred (time of day or day of the week) to identify "hot times." The corresponding predictive analytics methods start, at the most basic level, with regression analyses similar to what one would learn in an introductory statistics class and extend all the way to cutting-edge mathematical models that are the subjects of active research. Some methods also attempt to identify the factors driving crime risk.

Table 1.2 summarizes methods to identify individuals at high risk of offending in the future. The bulk of these methods relate to assessing individuals' risk. Here, conventional methods rely on clinical techniques that add up the number of risk factors to create an overall risk score. The corresponding predictive analytics methods use regression and classification models to associate the presence of risk factors with a percent chance that a person will offend. Also of interest are methods that identify criminal groups (especially gangs) that are likely to carry out violent assaults on each other in the near future. Hence, these methods can also be used to assess the risk that an individual will become a victim of crime.

---

[13] In this context, *large* refers to an amount of data beyond what a single analyst could recall without the assistance of a computer program or similar resources. Conversely, *low to moderate* refers to a data set that is sufficiently small that an analyst *could* reasonably recall its key facts.

Table 1.1
Law Enforcement Use of Predictive Technologies: Predicting Crimes

Problem	Conventional Crime Analysis (low to moderate data demand and complexity)	Predictive Analytics (large data demand and high complexity)
Identify areas at increased risk		
Using historical crime data	Crime mapping (hot spot identification)	Advanced hot spot identification models; risk terrain analysis
Using a range of additional data (e.g., 911 calls, economics)	Basic regression models created in a spreadsheet program	Regression, classification, and clustering models
Accounting for increased risk from a recent crime	Assumption of increased risk in areas immediately surrounding a recent crime	Near-repeat modeling
Determine *when* areas will be most at risk of crime	Graphing/mapping the frequency of crimes in a given area by time/date (or specific events)	Spatiotemporal analysis methods
Identify geographic features that increase the risk of crime	Finding locations with the greatest frequency of crime incidents and drawing inferences	Risk terrain analysis

Table 1.2
Law Enforcement Use of Predictive Technologies: Predicting Offenders

Problem	Conventional Crime Analysis (low to moderate data demand and complexity)	Predictive Analytics (large data demand and high complexity)
Find a high risk of a violent outbreak between criminal groups	Manual review of incoming gang/criminal intelligence reports	Near-repeat modeling (on recent intergroup violence)
Identify individuals who may become offenders:	Clinical instruments that summarize known risk factors	Regression and classification models using the risk factors
Probationers and parolees at greatest risk of reoffending		
Domestic violence cases with a high risk of injury or death		
Mental health patients at greatest risk of future criminal behavior or violence		

Table 1.3 summarizes methods used to identify likely perpetrators of past crimes. These approaches are essentially real-world versions of the board game Clue™: They use available information from crime scenes to link suspects to crimes, both directly and by processes of elimination. In conventional approaches, investigators and analysts do this largely by tracing these links manually, with assistance from simple database

Table 1.3
Law Enforcement Use of Predictive Technologies: Predicting Perpetrator Identities

Problem	Conventional Crime Analysis (low to moderate data demand and complexity)	Predictive Analytics (large data demand and high complexity)
Identify suspects using a victim's criminal history or other partial data (e.g., plate number)	Manually reviewing criminal intelligence reports and drawing inferences	Computer-assisted queries and analysis of intelligence and other databases
Determine which crimes are part of a series (i.e., most likely committed by the same perpetrator)	Crime linking (use a table to compare the attributes of crimes known to be in a series with other crimes)	Statistical modeling to perform crime linking
Find a perpetrator's most likely anchor point	Locating areas both near and between crimes in a series	Geographic profiling tools (to statistically infer most likely points)
Find suspects using sensor information around a crime scene (GPS tracking, license plate reader)	Manual requests and review of sensor data	Computer-assisted queries and analysis of sensor databases

queries (usually, the names, criminal records, and other information known about the suspects). Predictive analytics automate the linking, matching available "clues" to potential (and not previously identified) suspects across very large data sets.

Table 1.4 summarizes methods to identify groups—and, in some cases, individuals—who are likely to become victims of crime. These methods mirror those used to predict where and when crimes will occur, as well as some of the methods used to predict who is most likely to commit crimes. Predicting victims of crime requires identifying at-risk groups and individuals—for example, groups associated with various types of crime, individuals in proximity to at-risk locations, individuals at risk of victimization, and individuals at risk of domestic violence.

Prediction-Led Policing Processes and Practices

Prediction-led policing is not just about making predictions; rather, it is a *comprehensive business process*, of which predictive policing is a part. That process is summarized in Figure 1.1, which is loosely based on the Center for Problem-Oriented Policing's SARA (Scanning, Analysis, Response, and Assessment) model.[14] The first two steps are collecting and analyzing crime, incident, and offender data to produce predictions. Data from disparate sources in the community require some form of data fusion. Efforts to combine these data are often far from easy, however. The last two steps focus

---

[14] The SARA model is commonly used to solve crime problems in policing. For more information, see Center for Problem-Oriented Policing, "The SARA Model," web page, undated.

Table 1.4
Law Enforcement Use of Predictive Technologies: Predicting Crime Victims

Problem	Conventional Crime Analysis (low to moderate data demand and complexity)	Predictive Analytics (large data demand and high complexity)
Identify groups likely to be victims of various types of crime (vulnerable populations)	Crime mapping (identifying crime type hot spots)	Advanced models to identify crime types by hot spot; risk terrain analysis
Identify people directly affected by at-risk locations	Manually graphing or mapping most frequent crime sites and identifying people most likely to be at these locations	Advanced crime-mapping tools to generate crime locations and identify workers, residents, and others who frequent these locations
Identify people at risk for victimization (e.g., people engaged in high-risk criminal behavior)	Review of criminal records of individuals known to be engaged in repeated criminal activity	Advanced data mining techniques used on local and other accessible crime databases to identify repeat offenders at risk
Identify people at risk of domestic violence	Manual review of domestic disturbance incidents; people involved in such incidents are, by definition, at risk	Computer-assisted database queries of multiple databases to identify domestic and other disturbances involving local residents when in other jurisdictions

Figure 1.1
The Prediction-Led Policing Business Process

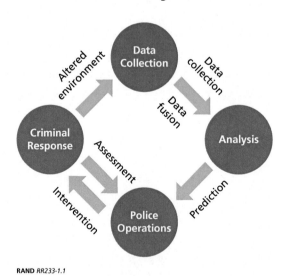

RAND *RR233-1.1*

on the response to the predictions. Police personnel use the predictions to inform their responses and then respond using evidence-based approaches. Criminals react to the changed environment. Some will be removed from the environment; those who are still operating may change their practices or move to a different area. Regardless of the response, the environment has been altered, the initial data will be out of date, and new data will need to be collected for analysis.

Data Collection

All predictive policing techniques depend on data. Both the volume and the quality of these data will determine the usefulness of any approach. The saying "garbage in, garbage out" strongly applies to these analyses. Efforts should be made to ensure that data are accurate and complete, though some techniques are less sensitive to small errors than others. Data sets will need to be updated periodically to ensure that they are current and reflect the effects of interventions. Well-trained analysts and researchers are also critical to predictive policing: Even with pristine data, a lack of strong analytics may result in less-than-desirable outcomes.

Data Fusion

The methods discussed in Chapters Two and Four rely heavily on data not only on crimes but also on the environment in which the crimes took place. Most crime data will likely be collected by police departments in the normal course of business, but information describing the environment may come from many other sources. Free and commercial data sets are available for use with crime data; examples of useful analytic additions include data on businesses, infrastructure, and demographics. As part of the data collection and analysis process, analysts must be able to combine disparate information sources. There are a number of methods for combining information, ranging from very simple techniques that offer an approximate picture (sometimes referred to as *heuristic* solutions) to more sophisticated methods that enable *information fusion*. In general, we found little evidence that departments have developed formal rules for combining data from disparate sources to form a cohesive picture of high-risk places, individuals, and groups. We address some specific technical methods for making sense of noisy and conflicting data in Chapter Four.

Analysis

Chapters Two through Four highlight different types of analytic methods for predicting crime and offenders. These methods should not be considered mutually exclusive; in fact, many of the techniques can be used in concert to provide more than the sum of their parts. Using regression and data mining techniques to explore available data sets can provide insights into crime patterns that are unique to a given region. The trends identified in this exploratory analysis can then inform the design of a method to identify hot spots. For example, these techniques can tell you how far back to look for crime patterns or whether there are seasonal or weekly trends that should be included in the analysis. GIS data mining can also be informed by regressions and can be used to explore data: Geographic profiles derived from clustering techniques can reveal patterns indicating a serial criminal.

Police Operations

Even the best analysis will do nothing to affect crime rates if it does not influence police practices. Locations identified as hot spots may require additional patrol attention, periodic visits by beat officers, or other responses that are appropriate for the types of crimes occurring there. Effective intervention must also occur in parallel to a continuous assessment process: To what extent has a police department's response reduced crime? An aggressive, objective assessment process is key to improving this response.

The type of intervention will vary with the situation and the department charged with intervening. Figure 1.2 shows three broad types of interventions, arranged (top to bottom) from simplest to most complex. In general, we hypothesize that the more complicated interventions will require more resources, but they will be better tailored to the actual crime problems—and get better results.

- *Generic interventions:* allocating more resources in response to increased risk. For hot spots, this might mean allocating more officers; for "hot people," this might mean allocating more parole or probation officer contacts.
- *Crime-specific interventions:* assigning resources that are tailored to combating the expected types of crime. Resources and interventions might focus on a given hot spot or a particular person who is at risk of offending.
- *Problem-specific interventions:* identifying location-, population-, or person-specific problems generating crime risk and fixing them. This level includes measures to investigate and solve specific crimes, almost by definition.

Regardless of the type of intervention, those carrying it out—from the command level to tactical officers—will need information to execute the intervention successfully. Thus, a critical part of any intervention plan is providing information that creates the needed *situational awareness* among officers and staff.

Figure 1.2
Intervention Methods

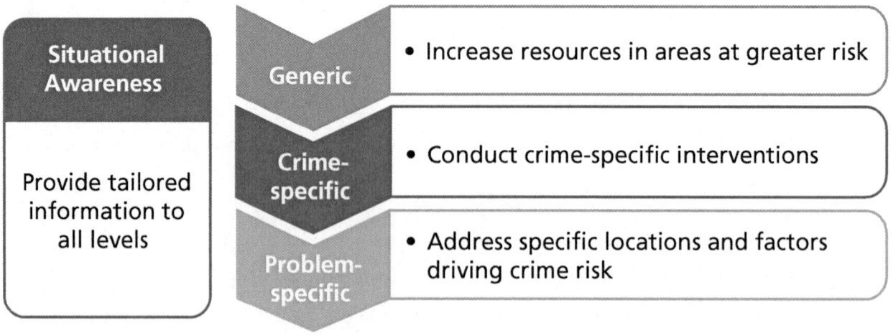

Situational Awareness	Generic	• Increase resources in areas at greater risk
Provide tailored information to all levels	Crime-specific	• Conduct crime-specific interventions
	Problem-specific	• Address specific locations and factors driving crime risk

RAND RR233-1.2

Criminal Response

Once the police launch an intervention, some criminals may be arrested and removed from the streets. Others may choose to stop committing crimes, change where they commit crimes, or change the way they go about committing crime in response to the police intervention. Thus, a location that had been hot can cool off, with some criminal activity moving to another area. These changes will make the original data set obsolete. In this way, the cycle begins again with a new round of data collection, analysis, and intervention.

About This Report

Chapter Two provides additional details on some promising methods used to predict the time and place of future crimes, as well as the victims of future crimes. That chapter concentrates on the first two steps of the prediction-led policing business process shown in Figure 1.1. Chapter Three focuses on intervention methods. Through several examples and case studies, it builds on the discussion in Chapters Two and Three by showing how police departments have worked through some or all four steps of the business process. In Chapter Four, we discuss the methods used to predict offenders and perpetrators of crimes. A section of that chapter also addresses concerns about individual civil liberties and privacy rights in the context of predictive policing. Finally, Chapter Five concludes this guide with findings and recommendations for practitioners, developers, and policymakers.

Making Predictions About Potential Crimes

When in doubt, predict the present trend will continue.

—"Merkin's Maxim"

In this chapter, we focus on predictions about crime and its victims: when and where it is most likely to occur, what is likely to cause it, and who is most likely to be a victim. The discussion focuses on the first two steps of the prediction-led policing business process depicted in Figure 1.1 in Chapter One. As with most forecasting methods, predicting future criminal events—whether from a tactical (next incident) or strategic (long-term) perspective—involves studying data on past crimes and victims, often using a variety of methods but generally always looking for patterns. The underlying assumption is that the past is prologue—at least the more recent past (days for tactical approaches, months to a few years for strategic approaches). Although this is true for addressing the questions of when, where, what, and who, the methods used will differ with the context and the goal:

- Hot spot analysis, statistical regression, data mining, and near-repeat methods are generally used to identify *where* a crime will occur over a specified time horizon (*when*, varying from a day to a year, depending on the method and application) and therefore *who* is likely to be a victim.
- Temporal and spatiotemporal methods can be used to identify *when* a crime is most likely to occur. These methods also identify victims (*who*) because they account for the ambient population, as well as local residents.
- Risk terrain analysis is appropriate for discerning the geospatial factors that create crime risk and looking for physical locations that might be ripe for a specific type of crime (*where*).

Practitioners should be very cautious when applying analytic methods aimed at answering the question, "What are the likely causes of the crime?" In general, they should always be cautious to avoid assigning causal relationships to the results of analyses using any of these methods. A statistical relationship between one factor and greater crime risk does not necessarily mean that the factor "causes" crime. A famous example illustrates this problem: Police tend to operate in areas with high crime, but it would be a mistake to say that police cause high crime. (It may or may not be the case that high crime leads to increased police operations.)

The predictive techniques to predict crime risk presented in this chapter are organized around six analytic categories: hot spot analysis, regression methods, data mining techniques, near-repeat methods, spatiotemporal analysis, and risk terrain analysis. Not all the techniques in these categories are equivalent in complexity. They are generally comparable only to the extent that all are used in some way or another to predict the time and place of future crimes, and they all depend on historical crime data.

We divide the techniques into four classes:

- *Classical statistical techniques:* This class includes standard statistical processes, such as most forms of regression, data mining, time-series analysis, and seasonality adjustments.
- *Simple methods:* Simple methods do not require much in the way of sophisticated computing or large amounts of data. Most heuristic methods, for example, are simple methods—relying more on checklists and indexes than on the analysis of large data sets.
- *Complex applications:* These applications include new and innovative methods or methods that require considerable amounts of data in addition to sophisticated computing tools. Many newer data mining methods and some near-repeat methods fall into this class.
- *Tailored methods:* In several cases examined here, existing techniques were adapted to support predictive policing. For example, classical statistical methods can be used to produce heat maps, which are simple, color-coded grids depicting the intensity of crime activity in a given area.

Table 2.1 matches each of the techniques addressed in this chapter with one or more of the four classes above and with each analytic category described earlier (when, where, what, and who).

Notes on Software

The software used for most of the methods discussed in this chapter is readily available. However, as software capabilities increase, so do the training requirements. Spreadsheet programs, such as Microsoft Excel and Calc in Apache OpenOffice (open-source and freely distributed), can do simple regressions and data manipulation, and they generally require little specialized training. For more complicated analysis or data manipulation, statistical software packages and languages, such as SAS, IBM's Statistical Package for Social Sciences (SPSS), and R (open-source and freely distributed), provide

Table 2.1
Classes of Predictive Techniques

Analytic Category and Primary Application	Predictive Technique	Class			
		Classical	Simple	Complex	Tailored
Hot spot analysis (*where*, using crime data only)	Grid mapping	X			X
	Covering ellipses	X			
	Kernel density	X			
	Heuristics		X		X
Regression methods (*where*, using a range of data)	Linear	X	X		
	Stepwise	X		X	
	Splines			X	X
	Leading indicators	X			X
Data mining (*where*, using a range of data)	Clustering	X		X	
	Classification	X		X	
Near-repeat (*where*, over next few days, using crime data only)	Self-exciting point process			X	
	ProMap			X	
	Heuristic		X		
Spatiotemporal analysis (*when*, using crime and temporal data)	Heat maps	X	X		X
	Additive model			X	
	Seasonality	X			
Risk terrain analysis (*where*, using geography associated with risk)	Geospatial predictive analysis			X	X
	Risk terrain modeling		X		X

access to very powerful techniques but also require significantly more training to use effectively.

Hot Spot Analysis and Crime Mapping

Hot spot methods predict areas of increased crime risk based on historical crime data. Hot spot methods seek to take advantage of the fact that crime is not uniformly distributed, identifying areas with the highest crime volumes or rates. The underlying assumption—and prediction—is that crime will likely occur where crime has already occurred: The past is prologue.

There are trade-offs when identifying hot spots, however. If the areas identified are too small, the results may exclude some areas of interest. But if they are too large, they may not be useful for allocating resources—and thus may not be "actionable." Very simple methods, such as grid mapping or thematic mapping, are commonly used, but these methods can be highly dependent on the initial data set and the partitioning of the map. Nevertheless, our discussion here begins by illustrating some simple approaches using grid mapping. We then focus on two more mathematically rigorous hot spot methods: covering ellipses and kernel density estimation. These two methods are relatively robust, are not sensitive to geographic partitioning, and have been incorporated into NIJ-sponsored analysis tools, including CrimeStat and HotSpot Detective.[1] All hot spot methods are related to clustering, which is discussed later in this chapter.

A critical consideration in predicting hot spots is how to act on this information—that is, what should be done with the results of the analysis? In a study to identify the optimal patrol duration to deter crime in hot spots, Christopher Koper used statistical analysis to determine how the likelihood of a crime being committed in a hot spot was affected by police patrols of varying durations.[2] Koper's research found a curve that related the length of a stop in a hot spot to a subsequent reduction in the likelihood of another crime being committed there. Patrol stops of 13–15 minutes, during which time police officers interacted with the members of the community, were found to be most effective. Thus, crime hot spot identification encourages more effective directed patrols.[3] More examples of taking action on predictions are provided in Chapter Three.

Grid Mapping

Figure 2.1 is a grid map of robberies in Washington, D.C. Squares colored red were in the top 2 percent for robbery counts (99th–98th percentile); orange squares were in the top 5 percent (97th–95th percentile); and yellow squares were in the top 5 percent (94th–90th percentile). Results shown are from a six-month weighted average of robbery data in 2009. Each cell is 250 meters long.

Qualitatively, the grid map has a good bit of "noise." The highlighted cells show clear patterns, but there is variation in where the colored cells are located, and isolated hot spots are common. It is not clear whether many of these isolated spots reflect a true

---

[1] Background, download instructions, and user information for CrimeStat III, version 3.3, are available at http://www.icpsr.umich.edu/CrimeStat; HotSpot Detective, version 2, is available for purchase at http://jratcliffe.net/hsd/index.htm.

[2] Christopher S. Koper, "Just Enough Police Presence: Reducing Crime and Disorderly Behavior by Optimizing Patrol Time in Crime Hot Spots," *Justice Quarterly*, Vol. 12, No. 4, 1995, pp. 649–672.

[3] Later in this chapter, we describe an intervention using the Koper model in Sacramento, California.

Figure 2.1
Robberies in Washington, D.C., by Weighted Grid Counts

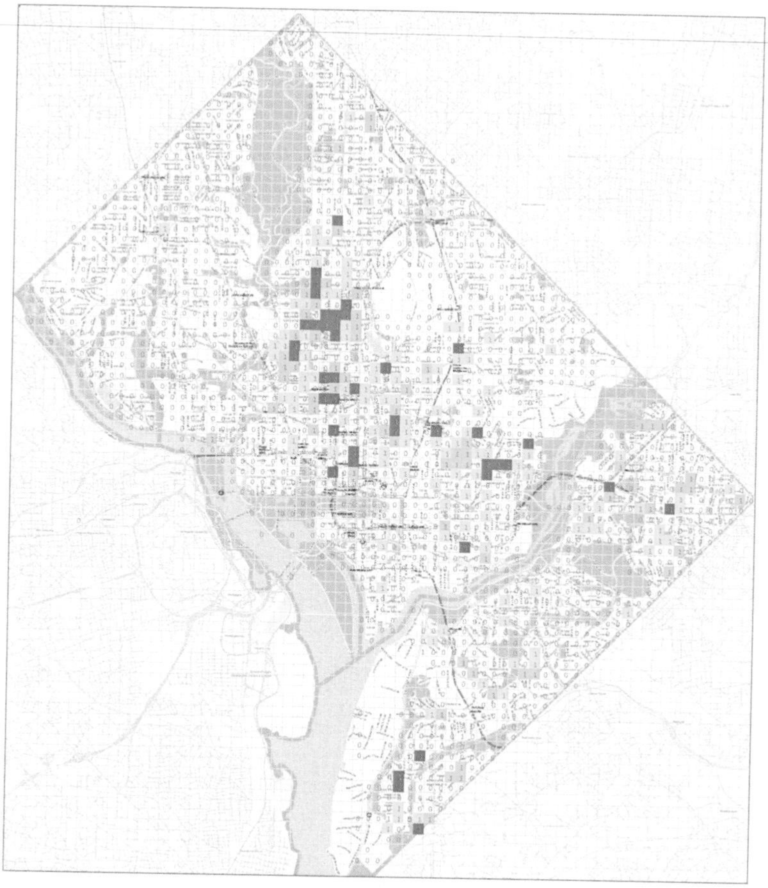

SOURCE: Preliminary findings from an RTI International, Structured Decisions
Corporation, and RAND project funded by NIJ's Office of Science and
Technology (OST).
RAND *RR233-2.1*

underlying crime risk or natural variation over the six months of data (i.e., "bad luck"
in those months).

Typically, grid maps are prepared using GIS programs, but this is not always necessary. Figure 2.1 was actually prepared using Excel.[4]

---

[4] For this analysis, robbery data from the Washington, D.C., Metropolitan Police Department were geocoded
using State Plane Coordinate System (SPCS) coordinates. These coordinates are rectilinear, with an easting (or
x-coordinate) and a northing (or y-coordinate). Given rectilinear data, it is straightforward to map each record
to a specific cell using Excel's floor function (i.e., floor, or easting/cell length, 1). From there, an analyst can use
Excel's pivot table function on the crime data to count the number of crimes in each cell. The "count" may be
a weighted sum or an average, if each record is given an appropriate weight. If the pivot table is set up to show
easting (x-axis) cell indexes on one axis and northing (y-axis) cell indexes on the other—and the cell heights and

Covering Ellipses

Hot spots can be identified by mapping crime instances and finding a set of ellipses that cover the clusters of occurrences. This technique is popular because it is easy to mathematically calculate the set of ellipses that encloses dense clusters of crime occurrences and because this functionality is readily available in the CrimeStat software.[5] Furthermore, because this method uses the geospatial position of each crime, it is not constrained by artificial boundaries when determining hot spots.

Often, the ellipses include a lot of area that is not actually high-crime because hot spots do not naturally form perfect ellipses. Several of the software implementations eliminate this problem by pruning the ellipses developed in the initial covering to their dense core. This two-step approach works by using the ellipses to cluster the crimes geographically and then defining the hot spots to be the smallest area that covers all crimes in a cluster (the convex hull of each cluster). This two-step approach has the advantage of identifying the most compact hot spots of significance, which ultimately better defines areas for directed patrols.

An example of covering ellipse methodology is nearest neighbor hierarchical clustering (Nnh). Nnh identifies groups of incidents that are spatially close. It is a hierarchical clustering routine that groups points together on the basis of a given criterion. The CrimeStat Nnh routine defines a threshold distance and compares the threshold to the distances for all pairs of points. Only points that are closer to one or more other points than the threshold distance are selected for clustering. Figure 2.2 illustrates the visual output from CrimeStat using Nnh to produce covering ellipses and, in this case, convex hulls.

While using the two-step approach will generally reveal in the densest hot spots of significance, the elliptical clustering method has two major deficiencies:

- These methods require the number of covering ellipses to be designated in advance, and it may not be immediately obvious how many are required. This problem can be mitigated by trying several different numbers of ellipses and choosing the "best" result. (Although the "best" result will be subjective, to a well-trained analyst, it will be clear and intuitive.)

---

widths are set to be equivalent—the result will be a grid map with weighted counts or averages in each cell. An analyst can then use Excel's conditional cell coloring functions to create cell colors like those shown in the figure. Copying the cells onto a map in Microsoft PowerPoint, and adjusting the size of the cells so that they line up with jurisdiction boundaries and landmarks, yields a rendering like Figure 2.1.

It is possible to take a similar approach in Excel with latitude-longitude coordinates. However, the formulas require trigonometric functions and can result in substantial spatial distortion over a large jurisdiction. Translating latitude-longitude coordinates to SPCS first is preferred.

5 Seven clustering methods are available to CrimeStat users and, in many cases, the program offers alternative algorithms for each method. They range from simple point location to more sophisticated, overlapping, and hierarchical techniques. However, in all cases, the output from each is represented as a set of ellipses depicting the hot spot areas.

Figure 2.2
Covering Ellipses and Convex Hulls Using Nearest Neighbor
Hierarchical Clustering

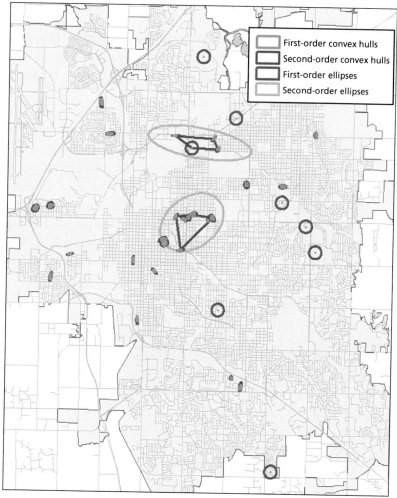

SOURCE: Susan C. Smith and Christopher W. Bruce, *CrimeStat III User Workbook*,
Washington, D.C., National Institute of Justice, July 2010, p. 56, Figure 5-10.
NOTE: This map is notional and does not represent an actual jurisdiction.
RAND RR233-2.2

- Because each observation is equally weighted, the final results can be very sensitive to the input data. Thus, if there are seasonal effects in the data, or if a very active residential burglar in a low-crime jurisdiction skews the statistics before being caught, crime data from three months ago may not be particularly relevant in identifying emerging hot spots. Thus, a data set covering six months can generate a very different set of hot spots than a smaller, more recent data set. This sensitivity can be useful, but it can lead to spurious patterns or trends and can make results difficult to reproduce. Therefore, special attention should be paid to ensuring that the data set contains only appropriate occurrences of crime.

Single and Dual Kernel Density Estimation

Kernel density estimation (KDE) is another approach for identifying hot spots. The intuitive idea is to spread out each crime's expected contribution to future crime risk over a certain area using a mathematical function called a kernel. KDE is a statistical analysis approach used to interpolate a continuous surface of crime data based on initial crime data points from different locations. The objective is to use crime incident data to identify hot spots based on their proximity to actual crime incidents. A kernel is a standardized weighting function used, in this application, to smooth crime incident data. In essence, users select a kernel function and a bandwidth and center the kernel function at the location of each crime incident. The area around each crime occurrence, as determined by the bandwidth, is weighted according to the kernel function.[6] Several kernel functions are available; CrimeStat offers the user five, each with its own advantages and disadvantages.[7] Most kernels are "peaked" in some way so that a crime's expected contribution to future crime risk diminishes with distance from that crime's location.

Single KDE estimates hot spots using a single variable: crime incidents. Dual KDE uses two variables, crime incidents and population density. In either case, the approach produces a contour map, a heat map, or a surface view map with the more heavily weighted areas of high crime visually represented. The hot spots can then be defined as areas above a certain threshold on each map.

Figure 2.3 illustrates a surface view and a contour view of a KDE interpolation of 1,180 street robberies in Baltimore County, Maryland, in 1996–1997. Both views were produced using CrimeStat, and both show three hot spot areas using the selected kernel and bandwidth.

A second example depicting the results of both a single and dual KDE is illustrated in Figure 2.4. Wiesenhütter and Oberwittler conducted a study of assault and battery incidents in the German city of Cologne using data from April 1999 to March 2000. They looked not only at the crime incidents but also at the population at risk in

---

6 A kernel function can be thought of as a distribution that is centered at a point (in this case, the location of the crime) and whose value indicates the relative influence the location has on the surrounding area. Some examples are uniform (flat), triangular (conical), standard normal (bell-shaped), and Epanechnikov (parabolic). CrimeStat features a negative exponential (or peaked) kernel function as well. See "Part III: Spatial Modeling" in the documentation accompanying the CrimeStat III software package, and T. J. Sullivan and Walter L. Perry, "Identifying Indicators of Chemical, Biological, Radiological, and Nuclear (CBRN) Weapons Development Activity in Sub-National Terrorist Groups," *Journal of the Operational Research Society*, Vol. 55, No. 4, April 2004.

7 Readers who are interested in a more mathematical discussion of KDE can refer to Walter Zucchini, *Applied Smoothing Techniques, Part 1: Kernel Density Estimation*, Philadelphia, Pa.: Temple University, October 2003.

Figure 2.3
Baltimore County Robberies, 1996–1997

Surface View   Contour View

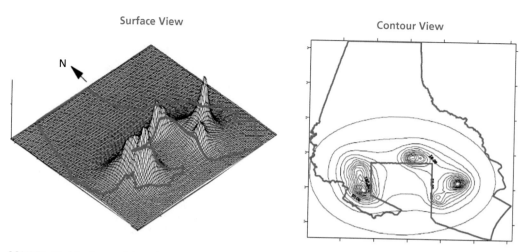

SOURCE: Ned Levine and Associates, *CrimeStat III: A Spatial Statistics Program for the Analysis of Crime Incident Locations*, prepared for the National Institute of Justice, November 2004, p. 8.18, Figure 8.9.
RAND *RR233-2.3*

Figure 2.4
Single and Dual Kernel Density Interpolation Results for Assault and Battery Incidents in Cologne, Germany

Single kernel density of assault
and battery incidents

Dual kernel density of assault and battery
incidents relative to at-risk population

SOURCE: Dietrich Oberwittler and Marc Wiesenhütter, "The Risk of Violent Incidents Relative to Population Density in Cologne Using the Dual Kernel Density Routine," in Ned Levine and Associates, 2004, p. 8.36.
RAND *RR233-2.4*

the various crime areas. For the incidents, they examined what they referred to as "calls to police," or the number of calls reporting assault and battery. For the population at risk, they examined the number of people spending time in the city, or the ambient population. To determine the ambient population, they calculated the number of people arriving and leaving through the city's 550 public transport stations.[8]

The single kernel density results indicate that the bulk of the crime incidents occur in the inner city. However, when the population at risk is taken into account in the dual kernel density interpolation, the risk to the population is much higher in several more distant areas that are known for their high concentrations of socially disadvantaged populations.

KDE is fairly robust and while the selection of the kernel and its bandwidth ultimately determines the nature of the hot spots, small changes in the function definition will result in small changes. For example, the results may change depending on the threshold used to define hot spots; small changes in the threshold should result in small changes to the hot spot distributions. Analyst experience with the data, along with an intimate knowledge of the jurisdiction, is crucial for appropriate selections when conducting KDE.

Because KDE does not start with an assumption about how many hot spots there should be, it avoids the first problem of elliptical covering. Like elliptical covering, KDE will be somewhat sensitive to the initial data set, but this can be mitigated for KDE by weighting the basis functions by their time or relevance to the period of the prediction.

This approach will result in less compact crime hot spots than will using ellipses followed by taking the convex hull. That may not be a significant problem, however, depending on the selection of a basis function. (Specifically, basis functions with smaller radii will result in crime hot spots that are more compact.)

In a 2008 paper, Chainey, Tompson, and Uhlig compared KDE to other hot spot mapping techniques, including elliptical covering (though without the convex hull step). They found that KDE resulted in the highest "prediction accuracy index" (PAI).[9] PAI, one proposed measure of effectiveness for hot spot methods, compares hit rates for crimes to the percentage of area that is predetermined to be "hot." On the strength of these results, it appears that KDE (single and dual) has demonstrated promise as a predictive tool for law enforcement.

Heuristic Methods

Interviews with crime analysts at several police departments in the United States and Canada revealed that, in many cases, predictive techniques consisted of purely heuris-

---

8   Oberwittler and Wiesenhütter, 2004.

9   Spencer Chainey, Lisa Tompson, and Sebastian Uhlig, "The Utility of Hotspot Mapping for Predicting Spatial Patterns of Crime," *Security Journal*, Vol. 21, No. 1, February 2008.

tic methods: Instead of using the more mathematically structured hot spot methods, departments often use less sophisticated techniques to uncover actionable crime hot spots. They found that these methods are effective because police departments know their cities. Specifically, they know the high crime neighborhoods, the types of crime likely to occur in each neighborhood, and when they are likely to occur. They know and understand the population and, generally, they can easily discern crime indicators.

We asked, "What predictive techniques do you get the most use out of?" One crime analyst responded by describing what can only be interpreted as purely heuristic tools:

> We have some tools for temporal analysis. Those reports are very useful, looking at time of day by week to see if there's a spike in particular crimes. We actually have in our roll-call rooms a widescreen TV where they can see what has occurred in the last 24 hours in the precinct, and an officer can get a good feel of what is happening in their area of responsibility but also what's taken place during other shifts. I think the crime maps help them a lot [as does] the year-to-year analysis. You can see if certain offenses are up or down. You can look at weekly trends compared against next year. If there was a high week in last year for burglaries, you can look at that and address those. Certain types of offenses have some seasonality to them, and you might see robberies pick up in November.[10]

With few exceptions, responses from other crime analysts reflected similar experiences. It appears that the objective is to interpret the results from sophisticated mathematical analyses, making it easier for police officers to access the information they need to understand the current situation and determine how to respond. One crime analyst put it this way: "The output from the dashboard is job-specific. It tries to give you the level of information that you need, given your role in the organization. We try to minimize the number of clicks."[11]

The following three methods are considered heuristic but were nevertheless cited as effective by the users with whom we spoke:

1. *Manual identification of hot spots on pin maps ("I eyeball it"):* Analysts often examine geocoded point data to identify hot spots. Using this approach, analysts use their judgment and experience to identify or define areas of concentrated activity. This approach is limited by the fact that repeat locations are often rendered as a single point, which skews any visual representation of a cluster. Despite obvious, nonscientific limitations, this approach is heavily used by both new and experienced analysts, and, often, when coupled with good jurisdictional knowledge, it is believed to be quite accurate.

---

[10] Department crime analyst, interview with the authors, August 2012.

[11] Department crime analyst, interview with the authors, August 2012.

2. *Quadrat thematic mapping (grid cells):* The manual version of the grid mapping methods discussed earlier, this technique involves aggregating crime data on a matrix of equally sized polygon cells or grids. The cells reflect "hot" areas, typically using shades of red or yellow.[12] We include this method here as heuristic because the color intensity in each cell is typically determined by the number of crime incidents recorded. However, the method is not restricted to counts; a smoothing function, such as a KDE, might be used to identify the color intensity as well. This approach is the same as the "simple density" option in ESRI's ArcMap, as opposed to a kernel density option. Figure 2.5 illustrates a quadrat thematic map of vehicle crimes using 250-meter quadrats. The analyst has manually singled out four areas of special interest because of the high crime rates depicted.

3. *Jurisdiction-bounded areas:* Analysts sometimes use a thematic map based on jurisdictional geography (e.g., districts, precincts, zip codes), typically using polygons to determine hot spots. Darker-colored polygons are often used to represent hot areas. On the positive side, this means that the polygons can correspond nicely to familiar areas (e.g., police beats). On the negative side, the use of jurisdictional or political boundaries subjects this approach to the "modifiable areal unit problem," which means that the selection of hot spots can reflect the geographies of

Figure 2.5
Thematic Quadrat Map of Vehicle Crimes

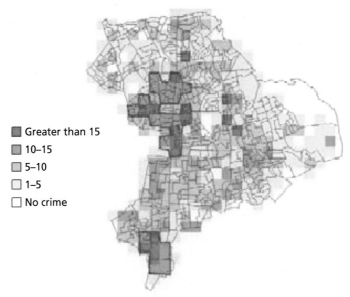

Greater than 15
10–15
5–10
1–5
No crime

SOURCE: Eck et al., 2005, p. 25, Exhibit 10.
RAND *RR233-2.5*

---

[12] Eck et al., 2005, pp. 25–26.

the polygons more so than a real underlying risk. Analysts can choose whether to use simple crime volume to indicate polygons of highest crime density or to base the coloring on a rate—for example, using population data to compare areas. Figure 2.6 shows a thematic map based on crime incidents in census tracts. The analyst has outlined three tracts that are considered to be particularly hot.

Regression Methods

Regressions fit a mathematical relationship between the variable to be predicted and independent "explanatory" variables. In contrast with hot spot mapping, regressions project future crime risk based not just on past crimes but also on what can be a wide range of data. For example, a regression model for future burglaries might include as inputs prior burglaries as well as counts for other types of crime, counts for vandalism and other types of disorder, numbers of homes in the area, numbers of unoccupied homes, the number of individuals with recent convictions for property crimes, and so on.

Statistical regression has been applied to problems related to crime for many years. Although more sophisticated techniques can provide additional information or

Figure 2.6
Crime Incidents, by Census Tract

SOURCE: Eck et al., 2005, p. 24, Exhibit 9.
RAND RR233-2.6

avoid some pitfalls, regression is still useful for many types of crime analyses. That said, regression predictions can be unstable if there is substantial volatility in the underlying data set or if the data set is small.

Regressions can be useful for questions to which the answer is a number with some confidence value associated with it. So, suppose the question is "How many robberies will a neighborhood have next week?" A regression might find the answer to be "There will most likely be eight, but we are 90-percent sure there will be between two and 14."[13]

A number of regression techniques can be used in predictive policing, ranging from variations in the kinds of mathematical relationships allowed to how analysts select which input variables and data sets to use.

Types of Relationships

In this section, we outline three relationships between the variables to be predicted and independent "explanatory" variables in regression methods:

- *Linear regression:* Linear regressions model the basic relationship between two or more variables by fitting a linear equation to observed data. This means that the equation is a weighted average of the input variables (e.g., predicting the expected number of robberies next month to be half the robberies committed last month plus one-fourth the disorderly conduct calls last month plus some constant number). The model is called *linear* because, geometrically, these methods fit a line or plane to the relationship between the output and input variables. These methods can also estimate confidence intervals on the crimes (e.g., "expected eight robberies with a 90-percent chance of between two and 14 robberies"). The most common method for fitting a regression line is the method of least-squares. This method calculates the best-fitting formula for the observed data by minimizing the sum of the squares of the deviations from each data point to the formula's prediction for that data point.[14]
- *Nonlinear regression:* These methods allow for more complicated mathematical formulas between the input and regression variables than weighted averages. Practically speaking, fitting nonlinear relationships requires more complex and time-consuming statistical algorithms.

---

[13] For a detailed discussion of sample size selection, see William G. Cochran, *Sampling Techniques*, 3rd ed., New York: John Wiley and Sons, 1977.

[14] Adapted from Yale University, Department of Statistics, "Linear Regression," supplemental course materials, September 16, 1997. Generating confidence intervals requires assuming that the predicted variable's values come from particular well-behaved random distributions. Continuing the example, say that the number of robberies in the neighborhood come from a normal distribution with a mean (average) equal to whatever the linear equation predicts.

- *Regression splines:* Regression splines allow different regression methods to model data over different regions of the dependent variable. For example, suppose that, for a given year, the probability of an auto theft in an area ranges from 0.01 percent to 45 percent. It may be that, between 0.01 percent and 5 percent, the best explanatory variables are model type, location of the vehicle, and age of the thief, whereas the best explanatory variable of thefts from 5 percent to 45 percent are the age and condition of the vehicle. Regression splines allow the analyst to create piecewise regressions over the full range of the dependent variable.[15]

Selecting Input Variables

Selecting which input variables to include in a regression model can be a challenge, especially if there is a large number of candidate input variables. While one can put all the possible variables into a model, the result will be a model that is *overfit*; the resulting formulas merely reflect random noise in the input data rather than a true relationship between the input and output variables. Some methods to deal with variable selection are as follows:

- *Manual experiments and correlation heuristics:* With these methods, analysts examine *correlations* (the extent of mathematical relationships) between the input and output variables and filter out input variables that are (1) not well correlated with the output variable to begin with or (2) too highly correlated with the output variable. The analyst adjusts the filtering criteria and otherwise experiments with different variables to find a model with good overall predictive accuracy and for which each input variable contributes meaningfully in a statistical sense.
- *Forward/stepwise regression:* These methods are "greedy" heuristics that iteratively build regression models. During each iteration, the method adds the additional input variable to the model that makes the statistically "best" improvement to the model's accuracy. (Stepwise methods can also drop input variables that cease to make statistically significant contributions to the model as new variables come in.) These methods have come under heavy criticism for being approximate and for giving the false impression that the variables making it into the final model "cause" the output variable (instead of just happening to be selected at a given iteration). These warnings aside, stepwise regression frequently generates very good predictive models, and the methods are broadly available in statistical packages and as Excel add-ins.[16]

---

[15] Adapted from Charles Nyce, *Predictive Analytics White Paper*, Malvern, Pa.: American Institute for Chartered Property Casualty Underwriters/Insurance Institute of America, 2007.

[16] Adapted from Nyce, 2007.

- *Mathematical optimization methods:* These methods place penalties on the number of variables included in the model and solve complicated optimization problems to fit the "best" overall model, with *best* having complicated definitions outside the scope of this guide. This class includes some of the most cutting-edge analytic models, including least-angle regression, lasso regression, and the elastic net.[17] Typically, one has to use very sophisticated tools to employ these techniques; note that most are available in the open-source package R. Although they are known for generating some of the most accurate models possible, they do have a downside: Since the goal is maximizing overall predictive power, the model parameters on any given input variable can be practically meaningless.

Leading Indicators in Regression (and Other) Models

A leading indicator is a sign that provides hints about what is to come. Just like a thunder cloud on the horizon indicates that rain may be coming, variables can indicate the direction that crime will be trending in the near future. These could be petty crimes that lead to more serious crimes, geographic changes that shift the locations of crime (such as construction of a mall, a sports arena, or a large building), or other changes that alter the location or intensity of criminal activity. Leading indicators allow analytical policing to move from reactive to proactive, but care should be taken to ensure the predictive models remain relevant. When some types or targets of crime cease to be profitable, the system may evolve in response, and new leading indicators will be needed.

Data on some leading indicators can be easily acquired. For example, weather has been found to lead to an increase in certain types of crime. Bushman et al. showed that higher temperatures correlate with higher crime, even when controlling for such factors as season and time of day.[18] Thus, the weather forecast could inform a statistical model determining where to anticipate crime in the very short term.

Gorr and Olligschlaeger compared different regression approaches for predicting a set of crime categories using data from Pittsburgh.[19] They ran regressions of different complexity on the same data set and compared the results. They found that simple time series were outperformed by more complicated methods. In particular, they found that because of some instability in the underlying data, using a smoothing coefficient (placing more weight on more recent data) improved predictions.

---

[17] Hui Zou and Trevor Hastie, "Regularization and Variable Selection via the Elastic Net," *Journal of the Royal Statistics Society: Series B (Statistical Methodology)*, Vol. 67, No. 2, April 2005.

[18] Brad J. Bushman, Morgan C. Wang, and Craig A. Anderson, "Is the Curve Relating Temperature to Aggression Linear or Curvilinear? Assaults and Temperature in Minneapolis Reexamined," *Journal of Personality and Social Psychology*, Vol. 89, No. 1, July 2005.

[19] Wilpen Gorr and Andreas Olligschlaeger, *Crime Hot Spot Forecasting: Modeling and Comparative Evaluation*, prepared for the U.S. Department of Justice, July 3, 2002.

More recently, Neill and Gorr collected two sets of crime offense reports from the Pittsburgh Bureau of Police. One set reported violent crimes, such as murder and armed robbery, and the other reported what the authors termed "leading indicator" crimes, such as simple assault and disorderly conduct. The detected violent crime clusters were used as the dependent variable, and the authors examined how many of these clusters could be predicted by the leading indicator data. Of the 93 significant violent crime clusters, 19 were successfully predicted by the leading indicator data.[20]

A Regression Example

Figure 2.7 revisits the forecasting of Washington, D.C., robberies discussed earlier in this chapter. This time, however, the cell colors show numbers of robberies forecasted from a regression model rather than weighted averages of robberies to date. The model makes predictions based on historical data within a cell and also within each of the eight neighboring cells (with edge-adjacent cells given higher weight than cells that are only corner-adjacent). The variables used in the model include not just robbery histories but also the histories of five other Part 1 crimes and leading indicators, including disorderly conduct and suspicious activity calls. The disorderly conduct and suspicious activity calls have been further categorized into separate types using text mining. The model itself is a linear regression model with the specific variables (specific crime histories and call-type histories) selected using stepwise selection.

Qualitatively, the result of the use of the model is to reduce the "noise" in Figure 2.1; the hot spots in the map that correspond to city landmarks, such as main shopping and entertainment districts, are now larger. A number of these hot spots have been labeled on the grid map.

However, a regression model like that shown in Figure 2.7 can go too far in "reducing noise." As shown in Figure 2.8, some seemingly isolated hot spots correspond to individual Metro (subway) stations or other comparatively small landmarks, which—since they see a great deal of foot traffic—are likely real hot spots rather than "bad luck." The analyst will need to look at maps and models using a variety of resolutions and methods, employing their own knowledge of the jurisdiction to get a thorough understanding of which apparent hot spots are genuinely high-risk and which are not.

Data Mining (Predictive Analytics)

The regression models introduced here are just one family of mathematical models for generating forecasts based on a set of input data. The generalization of the concept of

20 Daniel B. Neill and Wilpen L. Gorr, "Detecting and Preventing Emerging Epidemics of Crime," *Advances in Disease Surveillance*, Vol. 4, No. 13, 2007.

Figure 2.7
Robberies in Washington, D.C., by Multiple Regression

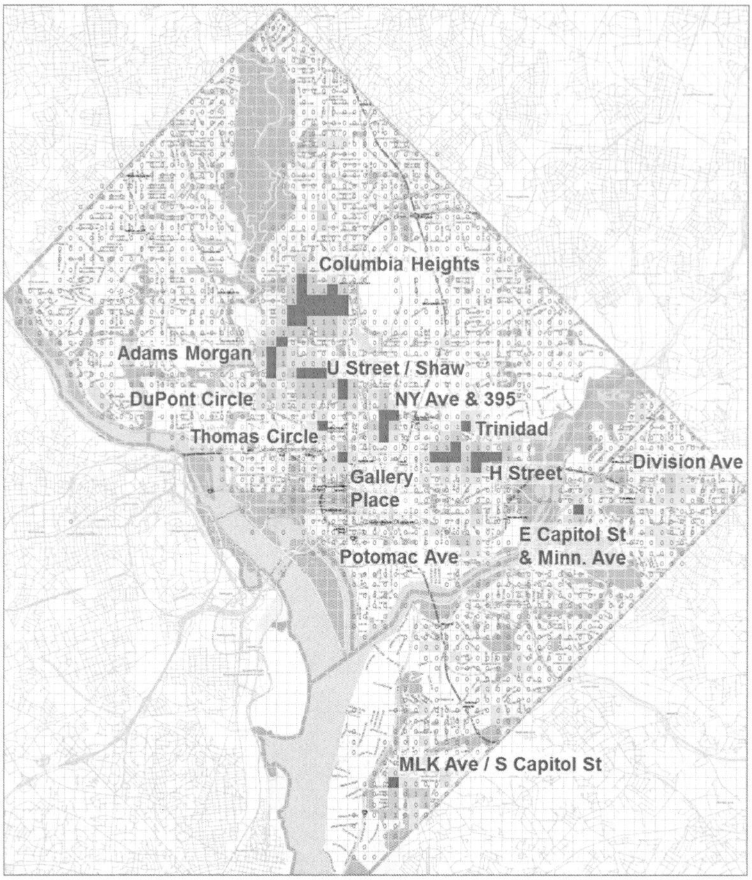

SOURCE: Preliminary findings from an RTI International, Structured Decisions Corporation, and RAND project funded by NIJ OST.
RAND *RR233-2.7*

building a mathematical model to make predictions based on input data is commonly referred to as *data mining*, or, in contemporary marketing, *predictive analytics*. Officially, data mining is the "practice of searching through large amounts of computerized data to find useful patterns and trends."[21]

Regression models are a subset of data mining, but the following methods also fall into this category:

---

[21] This definition is taken directly from the Merriam-Webster's online dictionary.

Figure 2.8
Robberies in Washington, D.C., with High-Risk Landmarks Highlighted

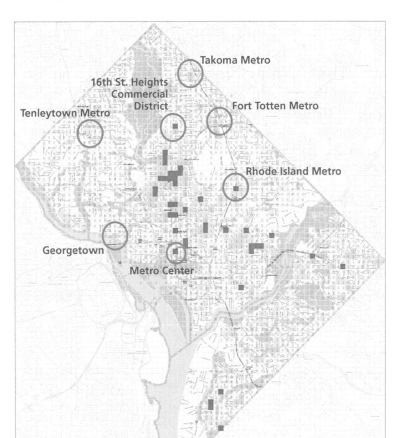

SOURCE: Preliminary findings from an RTI International, Structured Decisions
Corporation, and RAND project funded by NIJ OST.
RAND *RR233-2.8*

- *Classification* methods predict a category for an outcome (e.g., "There is an 85-
  percent chance of a robbery here next month"), rather than a continuous number,
  as in regression (e.g., "We predict an average of 1.24 robberies here next month").
- *Clustering* methods subdivide records into groups in which the records are "similar" mathematically. These models make predictions by stating that a future situation will likely be similar to a previous cluster of situations (e.g., "This neighborhood is showing attributes similar to those of other neighborhoods labeled as
  high-crime").
- There are methods that allow for far more complicated (or at least very different) relationships between input data and output predictions than is normal in
  regression models. Recall that regression models presume simple formulas relat-

ing input data and output relationships, such as "expected robberies = half of last month's robberies + half of last month's disorder-related calls + a random term."

- *Ensemble* methods take a number of simple predictive models and combine them in some way to yield a final overall prediction (such as by having the models vote or taking a weighted average of their output).
- In traditional regression analysis, the analyst typically works with only a handful of input variables to avoid computational problems. In many situations, the analyst suffers from a surplus of possible variables to choose from in building a predictive model—often hundreds, if not thousands. For example, in building a crime forecasting model, the analyst might have data on multiple types of prior crime, multiple types of disorder reports, multiple economic and demographic variables, as well as weather and data, and seemingly endless ways of representing or combining them. The field of data mining includes methods specifically designed to deal with situations in which there are many input variables.

The use of data mining models beyond linear regression has become more popular as computers become faster and are able to process more data in a short period. There is no single core technique for data mining; instead, data mining should be thought of as a suite of tools to extract information from large amounts of data. Indeed, it is fairly common to experiment with a number of algorithms and parameters, selecting the modeling approach that performs best.

The specific computational technique used in data mining depends on the nature of the data being analyzed. Therefore, the first step is to understand the data. There are different techniques depending on whether the dependent variable (frequently called the target variable or function) is *discrete* or *continuous*. As described here, a discrete variable is one that can take only a few sets of values (e.g., race, sex, or other categorical value). A continuous variable in this case will be any numerical value that can take any value within a range (e.g., property value, temperature).

One also can choose techniques of varying complexity. At one extreme, there are simple methods that select a relative handful of the possible variables and build a mathematical model with fairly simple relationships between the input data and the forecast. These models run comparatively quickly, and the results are usually directly interpretable by a person. Most regression analyses fit into this category, as do decision-tree methods (which generate a simple tree of "if-then" clauses to make a prediction). At the other extreme are methods that take most if not all of the possible variables and combine them using extremely complicated relationships to generate forecasts. These models run comparatively slowly, and the results are usually not interpretable by a person, except perhaps through a limited readout stating that certain variables had heavier or lighter "weights" in the final model. Such models are commonly referred to as *black box* models.

The advantage of these complicated methods is that they can *sometimes* produce forecasts that are more accurate than those generated by the simpler methods. For some models, however, drawbacks could include a loss of interpretability and much more significant computation requirements (though the computational cost depends on the algorithm).[22] Computation time can be a minor annoyance when working with a few hundred data points, but it is a potential project-ender when dealing with very large data sets (hundreds of thousands to millions of records).

Further, and not surprisingly, some techniques do not work if parts of the data are censored, observations are missing, or errors are present. Other methods are *robust* to these sorts of data problems—the accuracy of the predictions merely degrade as errors increase.

While we do provide a summary of the key types of data mining methods in the following sections, this guide merely scratches the surface of data mining. A thorough and highly readable discussion of the spectrum of data mining techniques can be found in the book *Data Mining: Practical Machine Learning Tools and Techniques* by Ian Witten and Eibe Frank.[23] For those seeking online instruction, Carnegie Mellon University's Auton Lab maintains a popular set of tutorials on a wide range of data mining techniques.[24]

Many of the software tools mentioned in the section "Regression Methods" are also capable of data mining, but there are additional tools designed strictly for this purpose. One such tool is WEKA. Developed by Witten and Frank, WEKA is a freely available, open-source program with an intuitive graphical interface that allows users to explore a wide array of data mining techniques.[25]

There are three families of data mining techniques that will be of primary interest for predictive policing: regression, clustering, and classification. Regression methods were introduced earlier; here, we just mention that there are techniques within this family that are much more complicated than the basic methods discussed in this guide. To choose among numerous input variables, one can, for example, use an algorithm that solves a complex optimization problem to determine what weighted combination of variables yields the best performance in a regression.[26] As mentioned already, clus-

---

[22] On the other hand, some complex methods are designed to accommodate extremely complicated models with low computational requirements. Examples are reproducing kernel methods and boosting.

[23] Ian H. Witten and Eibe Frank, *Data Mining: Practical Machine Learning Tools and Techniques*, 2nd ed., Burlington, Mass.: Morgan Kaufmann, 2005.

[24] Andrew Moore, "Statistical Data Mining Tutorials," Auton Lab, Carnegie Mellon University, 2006.

[25] Machine Learning Group, University of Waikato, "Weka 3: Data Mining Software in Java," homepage, undated.

[26] The Elastic Net algorithm is one of the more powerful methods for this purpose. See Zou and Hastie, 2005. Elastic Net is available through the statistical system R in the glmnet package. For more information, see Jerome Friedman, Trevor Hastie, and Rob Tibshirani, "glmnet: Lasso and Elastic-Net Regularized Generalized Linear Models," Comprehensive R Archive Network, undated.

tering approaches group data records into similar groups. And similarly, classification techniques assign data records to categories (commonly referred to as *classes*).

Clustering

Clustering algorithms form a class of data mining approaches that seek to group data into clusters with similar attributes. The goal is to find clusters of observations in which the observations in a cluster are significantly more similar to each other than to those outside of the cluster.

Clustering will work well with multiattribute data. For a burglary, for example, an observation might include such attributes as the neighborhood, time of day, entry method, or types of objects stolen. The clustering method selected will depend on whether the data are categorical or numerical, but most software implementations can automatically select the appropriate technique. Most clustering algorithms run quickly and can work with very large data sets.

Clustering algorithms can be used as part of a data exploration process to find commonalities across crimes. For example, clustering could be used on a data set of burglaries to find those exhibiting similar tactics. This could be evidence of a serial burglar, or it could suggest interventions to use against some of the clusters. As a specific example, Adderley and Musgrove cluster perpetrators of serious sexual assaults using a technique called a self-organizing map (SOM).[27] The SOM clusters observations with similar traits to form related groups. The authors were able to identify crimes committed by the same individual because of these similarities. This information could be fed into the hot spot models discussed earlier to identify the perpetrator's next target, or it could be used with the geographic profiling methods discussed in Chapter Three to identify the perpetrator's home base.

As another example, spatial clustering algorithms running on geospatial crime incident data can find statistically significant hot spots. *Indicators of spatial association* are statistics used to evaluate the existence of spatial clusters for a given variable. One example is local indicators of spatial association (LISA).[28] LISA statistics, such as Getis-Ord Gi*, compare local values to global values to determine the extent of spatial clustering. The Gi* statistic identifies areas that are more clustered than what would be expected by chance. The output—a z-score—indicates the magnitude of the

---

[27] Richard Adderley and Peter B. Musgrove, "Data Mining Case Study: Modeling the Behavior of Offenders Who Commit Serious Sexual Assaults," *Proceedings of the 7th ACM SIGKDD International Conference on Knowledge Discovery and Data Mining*, New York: Association for Computing Machinery, 2001.

[28] Luc Anselin, *Local Indicators of Spatial Association (LISA)*, Regional Research Institute Research Paper No. 9331, Morgantown, W. Va.: West Virginia University, 1994; J. K. Ord and Arthur Getis, "Local Spatial Autocorrelation Statistics: Distributional Issues and an Application," *Geographical Analysis*, Vol. 27, No. 4, October 2005.

clustering. Higher or lower z-scores reflect a greater degree of clustering, while scores near zero indicate spatial randomness.[29]

Classification

Classification algorithms seek to establish rules assigning a class or label to events. Classification techniques can be applied to any problem for which the analyst can use a categorical prediction rather than a numerical one.[30] Further, some classification techniques can find complicated nonlinear patterns relating input data to predictions.

Classification works with a training data set to learn the patterns that determine the class of an observation so that the pattern can be used to make predictions about future observations. For example, classification might be used to figure out which gang members about to be released are likely to reoffend based on data about previous releases of gang members. Similarly, classification methods might be used to predict whether a given area will be "high-crime," "medium-crime," or "low-crime" over the next month. For some classification methods, the techniques can generate a predicted number (e.g., an anticipated average robbery count), given that a record has been assigned to a particular category (e.g., the "high-risk" category).

As mentioned, these data mining techniques include comparatively simple methods, such as decision-tree algorithms, which run quickly and produce human-readable outputs—decision trees. A tree might look like this:

- IF a robbery happened here last month, predict "high robbery risk."
- ELSE:
  - IF there were more than five disorderly conduct calls, predict "medium robbery risk."
  - ELSE predict "low robbery risk."

Conversely, data mining also includes some of the most complex black box methods, which combine all variables into complicated formulas to make predictions. Examples of these more complicated algorithms include the *neural network* and *support vector machine* families, both of which are outside the scope of this guide.

Also popular are "ensemble" methods that generate a large set of simple predictive models, then combine them in some way (e.g., by having the simple models vote, averaging the models' predictions, or using the models' results as input data in a

---

[29] For statistically significant positive z-scores, the larger the z-score, the more intense the clustering of high values. For statistically significant negative z scores, the smaller the z-score, the more intense the clustering of low values.

[30] The category can be a number. For example, a classification model predicting that the number of robberies next month will most likely be close to "0," "2," "4," or "8" could use these category designations.

supermodel). The following are some of the more sophisticated and popular ensemble methods:

- *Boosting methods* start with an initial simple classification model (which is probably weak, or not terribly accurate) and iteratively add simple classification models to it. The idea is that the combination of simple models will be more accurate. For each iteration, boosting methods focus on finding a simple model that will correctly classify cases that were incorrectly classified in the last iteration. Depending on the method, the final prediction is either a weighted average or majority vote of all the simple classification models.[31]
- *Random forest methods* iteratively generate a large number of simple decision trees (a "forest"). In each iteration, they "grow" a simple decision tree on randomly selected subsets of input variables and input data. The final prediction is the plurality vote of all the simple trees.[32]
- *Bagging methods* generate a large number of classifiers—this time by building a classification model in each iteration on a randomly selected subset of the input data. The classifiers are typically decision trees or neural nets (with no requirements to be "simple"). Again, the final prediction is either the average or plurality vote.[33]

These more complicated methods do offer the potential of improved accuracy, though it is far from guaranteed that they will perform much better on any particular problem.[34] They also usually have good running times, as the algorithms operate by chaining together a number of simple models, each of which is generated quickly. The analyst may want to experiment with simple, black box, and ensemble methods (assuming that trying certain methods will not crash the analyst's computer) to see whether the increase in accuracy is worth the time and loss of understandability.

---

[31] See, for example, Robert E. Schapire, "The Boosting Approach to Machine Learning: An Overview," in David D. Denison, Mark H. Hansen, Christopher C. Holmes, Bani Mallic, and Bin Yu, eds., *Lecture Notes in Statistics: Nonlinear Estimation and Classification*, Vol. 171, New York: Springer, 2003.

[32] See Leo Breiman, "Random Forests," *Machine Learning*, Vol. 45, No. 1, October 2001.

[33] See Leo Breiman, "Bagging Predictors," *Machine Learning*, Vol. 24, No. 2, August 1996.

[34] For example, Caruana and Niculescu-Mizil conducted an empirical comparison of ten families of data mining algorithms across 11 binary classification data sets. They found that the ensemble methods (boosting on decision trees, random forests, bagging on decision trees) performed best overall, followed by support vector machines and neural networks, with the simpler methods at the bottom (nearest neighbors, single decision tree, regression, and naïve Bayes classifier). However, the data sets and problems used varied widely, including predicting personal income, recognizing letters, and determining the prevalent type of tree in an area. As a consequence, the algorithms used for particular problems also varied widely. See Rich Caruana and Alexandru Niculescu-Mizil, "An Empirical Comparison of Supervised Learning Algorithms Using Different Performance Metrics," *Proceedings of the 23rd International Conference on Machine Learning*, New York: Association for Computing Machinery, 2006.

As an example classification application, Sullivan and Perry studied the development cycle of terrorists seeking to acquire weapons of mass destruction and identified some indicators for the path of a terrorist organization using statistical classification techniques.[35] Specifically, they used a classification tree method and another method called discriminant analysis because their model had a discrete dependent variable (the organization's stage of weapon development: nonseeking, seeking, or strongly seeking). The authors found that organizational characteristics, such as the leadership's mindset and the group's technical capabilities, were strong predictors of acquiring weapons of mass destruction. Similar techniques can be used to identify factors that influence crime trends. The structure of the study could be duplicated to identify leading indicators to be fed into a regression model.

Training and Testing a Model

It is worth noting that verifying the accuracy of a predictive model is usually a two-step process. Especially if there are many input variables and possible ways of combining them, there is some risk that the predictive model "fit" in a data mining method is *overtrained*, meaning that the model reflects natural noise in the input data more so than genuine statistical relationships between the input data and output predictions. To resolve this, it is common practice to split the data set into a *training set* and a *testing set*. (The output variable is "known" for each set.) The data mining method fits the model to the training set; the model is then applied to the test set to see whether it has about the same predictive accuracy as on the training set. This will give the user a fair notion of the accuracy of the model generated by the data mining method. While most commonly associated with the more complex data mining models, this technique is also used to test the goodness of simpler models, including regression models.

Near-Repeat Methods

Near-repeat methods operate on the assumption that some future crimes will occur very near to current crimes in time and place—that areas recently seeing higher levels of crime will see higher crime nearby in the immediate future. George Mohler at Santa Clara University has documented several studies that support this assumption. For example,

- Burglars repeatedly attack clusters of nearby targets because local vulnerabilities are well known to the offenders.

---

[35] Sullivan and Perry, 2004.

- A gang shooting may incite waves of retaliatory violence in a rival gang's territory.[36]

Mohler claims that crime spreads through local environments (micro-time and micro-place) much like a contagious disease. If there has just been a crime, the risk of crime increases a short distance away, for a short period of time. This seems to be particularly true of burglaries. For example, in the San Fernando valley of Los Angeles in 2001–2005, there were more than 100 burglaries within three hours and 200 meters of a prior burglary.

Other studies have confirmed this pattern. For example, Michael Townsley and others have reported that the repeat rate (the proportion of repeat incidents that make up the overall crime count) was 18.7 percent in Beenleigh, Australia, the city with the sixth highest residential burglary rate in the state of Queensland.[37] They also found that the chance of a residential address being victimized one time only was 7 percent. Having been a victim once, the chance of revictimization was more than 16 percent— more than double the initial chance of becoming a victim.

To take advantage of this effect, Mohler et al. developed a "self-exciting process," or "earthquake modeling" algorithm.[38] This algorithm has been widely reported on by the media, as discussed in Chapter One. Although this algorithm uses sophisticated mathematics, the approach is relatively simple:

- Lay a grid over the jurisdiction, as in traditional grid mapping.
- Estimate the current rate (called the "background rate") at which new crimes appear in each grid square. This rate depends only on the characteristics of each grid square and jurisdiction-wide temporal effects. (For example, we can allow the current rate to increase in the summer months, when crimes are more frequent.)
- When there has just been a crime, assume that the rate for new crimes will jump up temporarily. (This can be thought of as the "aftershock rate.") This jump declines the longer the grid square goes without seeing a new crime, eventually falling back to the background rate.

Unsurprisingly, the "earthquake modeling" label comes from the fact that similar models are used to account for the likelihood of earthquakes before and immediately

---

[36] G. O. Mohler, M. B. Short, P. J. Brantingham, F. P. Schoenberg, and G. E. Tita, "Self-Exciting Point Process Modeling of Crime," *Journal of the American Statistical Association*, Vol. 106, No. 493, 2011.

[37] Michael Townsley, Ross Homel, and Janet Chaseling, "Repeat Burglary Victimisation: Spatial and Temporal Patterns," *Australian and New Zealand Journal of Criminology*, Vol. 33, No. 1, April 2000.

[38] Mohler et al., 2011.

after other earthquakes. A significant number of calculations, including simulation, are required to estimate the background and aftershock rates for all the grid squares.[39]

As an example application, the Santa Clara County Sheriff's Office in California implemented a version of Mohler's self-exciting point process with a grant from the U.S. Department of Justice. The focus was on property crimes (burglaries of homes and vehicles). Results from 2010 and 2011 were encouraging:

- The most common times for such crimes were Tuesdays and Thursdays between 5:00 p.m. and 8:00 p.m. The department provided officers with hot spot locations and victim profiles, and it claims that crime rates decreased.
- From 2010 to 2011, the property crime rate in the West Valley patrol area decreased by 23 percent.[40]

The Santa Cruz Police Department, also in California, implemented a version of Mohler's algorithm on a six-month trial basis beginning in July 2011.[41] After the program proved to be very successful, the department moved to full operation with the algorithm in July 2012. Comparing crime statistics for the first six months of 2012 with statistics for the same period in 2011, the department reported that property thefts were down 19 percent without additional officers or shifts.[42]

A near-repeat calculator has been developed by Jerry Ratcliffe at Temple University with a grant from the NIJ and is freely available.[43] The calculator software is based on the concept of "communicability" of risk to nearby locations for a short amount of time. The method builds on a space-time clustering methods first pioneered by George Knox in 1964. Knox studied the epidemiology of childhood leukemia. His test sought to determine whether more event-pairs could be observed with a closer proximity in space and time than would be expected under random distribution. The test used

---

[39] The details are outside the scope of this guide but can be found in Mohler et al., 2011. A very rough summary is as follows: At each iteration, the algorithm starts with a matrix of spatiotemporal distances between each crime and a matrix **P**, in which each entry is a probability that crime i "caused" crime j (i.e., it is an aftershock crime rather than a background crime). The algorithm then uses a random sample of crime point-pairs to develop estimates of the background and aftershock rate parameters using kernel density estimation techniques. Those new parameters are then used to update **P**, and the algorithm iterates until the rate parameters and **P** converge. The algorithm takes advantage of the fact that there are simple algebraic relationships between the true crime rate parameters and the true values of the matrix **P**.

[40] These statistics were reported in Josh Koehn, "Algorithmic Crime Fighting," *SanJose.com*, February 22, 2012.

[41] Kalee Thompson, "The Santa Cruz Experiment: Can a City's Crime Be Predicted?" *Popular Science*, November 1, 2011.

[42] Brian Heaton, "Predictive Policing a Success in Santa Cruz, Calif.," *Government Technology*, October 8, 2012.

[43] Jerry H. Ratcliffe, "Near Repeat Calculator," version 1.3, Philadelphia, Pa., and Washington, D.C.: Temple University and National Institute of Justice, August 2009. The following discussion draws on Cory Haberman and Jerry H. Ratcliffe, "The Predictive Policing Challenges of Near Repeat Armed Street Robberies," presentation to the International Crime and Intelligence Analysis Conference, Manchester, UK, November 3–4, 2011.

Monte Carlo simulation to compare actual time differences between events to time differences chosen randomly to determine whether time differences tend to be shorter than would be the case if events occurred randomly in time.

Another near-repeat method for predicting burglaries is ProMap.[44] This method assesses which grid squares are predicted to have the greatest imminent risk for burglaries based on where burglaries have recently been and uses a simple mathematical model. ProMap outperformed both kernel density maps and analyst-generated maps in an experiment in Merseyside, UK.

It is important to note that there is a simple heuristic version of near-repeat methods: Mark areas that have just seen a crime, especially a burglary, as being at elevated risk of another similar crime in the near future. This technique appears to work best with burglaries; other types of crimes do not show near-repeat effects that are quite as strong.

Spatiotemporal Analysis

The predictive methods discussed to this point have focused on the crime incident: crime type, location, and time, or the current patterns of crime. We know that crime patterns can change over time, however. In this section, we expand the discussion to include the relationship between crime and the environment over time. Here, we expand the prediction problem to include various environmental and temporal features of the crime location. The idea is to use this information, along with crime incident information, to predict the location and time of future crimes.

Basics of Spatiotemporal Analysis

Heuristic methods used to analyze spatiotemporal features are rather common. Generally, this is a manual process, perhaps with some help from a graphical user interface. The process can also take into account such information as food stamp distribution dates and data on the release of serial offenders. However, more mathematically rigorous methods have been developed to combine these features with information about the crime itself, as we discuss later in this section. Features generally used in spatiotemporal analysis include

- time of day, day of week, and time and day cycles
- temporal proximity to other events (e.g., payday, sporting events, concerts)
- season

---

[44] Shane D. Johnson, Kate J. Bowers, Dan J. Birks, and Ken Pease, "Predictive Mapping of Crime by ProMap: Accuracy, Units of Analysis, and the Environmental Backcloth," in David Weisburd, Wim Bernasco, and Gerben J. N. Bruinsma, eds., *Putting Crime in Its Place: Units of Analysis in Geographic Criminology*, New York: Springer, 2009.

- weather
- interval between offenses in a crime series (including correlations of those intervals to other factors, such as the value of stolen property)
- repeat locations
- geographic progression of incidents in a crime series
- spatial arrangement of incidents
- type of location (e.g., parks, convenience stores, public housing)
- geographic correlates (e.g., near bus stops, near establishments licensed to sell liquor)
- environmental and target factors (e.g., lighting, neighborhood condition, traffic level)
- demographic and economic data from the crime area.

All of these features—alone or in combination—have predictive value in analyzing both short-term series and long-term problems or hot spots. For example, analysts often find that serial offenders "space" their crimes in predictable intervals while adjusting to their preferred day of the week. An analyst might find a strong correlation between the cash stolen in bank robberies and the number of days the offender waits until the next offense. A forecast of snow might herald an increase in traffic collisions at specific intersections, especially if it is expected to occur during specific time frames.

Considerations of time and space apply not only to crime locations (as generally extracted from police data systems) but also to other locations related to the crime. For instance, a sexual assault or robbery might involve a location and time at which the offender "acquires" a victim that is separate from the location where the actual robbery or assault occurs. A vehicle might be stolen in one area, driven to another area for use in a crime, and dumped in a third area. These associated areas often have greater predictive value than the central crime location.

Crime analysts often make predictions or "forecasts" from these features using basic descriptive, inferential, and bivariate statistics. For example, a popular corporate training course encourages analysts to take the means and standard deviations of the intervals between offenses and use these statistics to create a predicted "window" for next events.[45] The IACA's Professional Training Series teaches analysts to (among other things) forecast crimes temporally using a simple linear regression between intervals and dollar values.[46] A spatial statistics training manual for analysts instructs on spatial forecasting by considering an offender's spatiotemporal moving average over time.[47]

---

[45] See for example, Steven Gottlieb, Sheldon Arenberg, and Raj Singh, *Crime Analysis: From First Report to Final Arrest*, Montclair, Calif.: Alpha Group Center for Crime and Intelligence Analysis Training, 1994.

[46] Samantha L. Gwinn, Christopher Bruce, Julie P. Cooper, and Steven Hick, eds., *Exploring Crime Analysis: Readings on Essential Skills*, 2nd ed., Overland Park, Kan.: International Association of Crime Analysts, 2008.

[47] Smith and Bruce, 2010.

Several other "predictive" techniques, particularly with large data sets, involve simple considerations of frequency distribution and modes: Locations of future hot spots and times for robberies are presumed to be the same as locations of past hot spots and times for robberies, perhaps adjusted slightly by known upcoming changes in the geography, demographics, or economics of the jurisdiction.

Heat Maps

Perhaps the simplest way to conduct spatiotemporal analysis is a *heat map*, a table that shows, through color intensity, the relative frequencies of crimes with different dates, times, and conditions. The heat maps shown in Figure 2.9 were prepared in Microsoft Excel using a database of more than 87,000 Part 1 crime records from the Washington, D.C., Metropolitan Police Department (incidents studied were from July 2007 to November 2009).

Preparing the heat maps involved simply creating new variables for the hour and day of the weeks in which the crimes reportedly occurred. We used pivot table features in Excel to create the tables showing the numbers of crimes by hour and day of the week. Then, we used conditional formatting features to color the cells. A variety of other packages offer similar heat map functionalities as well.

The heat maps show differing patterns for burglaries and robberies. Burglaries are concentrated during the daytime of the workweek (Monday–Friday) and are concentrated in the morning, particularly in the 7:00 a.m. hour. The 7:00 a.m. concentration is probably due largely to the timing of data reporting, reflecting when the property owner or manager discovered the burglary as opposed to when it occurred. Conversely, robberies tended to occur at night and were concentrated between 8:00 p.m. and midnight during the workweek and between 9:00 p.m. and 4:00 a.m. on weekend nights.

The heat map shown simply reflects time of day and day of week for three crime categories in all of Washington, D.C., but similar maps might be generated to show

- crime levels for particular districts, beats, or other jurisdictional boundaries
- crime levels by month
- crime levels during particular holidays
- crime levels during special events (e.g., sporting events, major expos)
- crime levels by weather conditions (assuming the data include a field for weather conditions in that jurisdiction at that time).

Beyond Excel, dedicated predictive policing tools can support spatiotemporal analysis. Figure 2.10 is a screenshot from Information Builders' Law Enforcement Application software. As shown, the analyst can identify likely hot spots on a city map (Richmond, Virginia, is shown here) for an upcoming date and time (four-hour block) under varying weather conditions.

Figure 2.9
Heat Map of Part 1 Crimes, Burglaries, and Robberies in Washington, D.C.

Six Part 1 Crimes

Hour	Sun	Mon	Tue	Wed	Thu	Fri	Sat
12:00 a.m.	829	483	464	421	480	546	874
1:00	634	284	298	278	297	321	677
2:00	534	223	213	202	231	243	535
3:00	477	168	140	133	141	173	472
4:00	334	163	156	155	141	169	317
5:00	136	100	104	107	94	112	115
6:00	140	240	241	208	220	210	214
7:00	275	612	602	574	612	600	342
8:00	183	390	358	388	402	393	192
9:00	313	609	556	540	524	547	380
10:00	413	486	483	424	462	480	434
11:00	424	485	463	473	475	498	483
12:00 p.m.	602	603	614	600	587	658	593
1:00	455	530	468	468	530	534	506
2:00	567	587	523	560	503	611	622
3:00	603	556	589	606	582	699	673
4:00	582	607	621	669	559	744	654
5:00	599	740	634	665	649	889	657
6:00	640	851	753	767	754	932	723
7:00	657	790	781	754	754	871	715
8:00	691	766	710	729	703	822	721
9:00	665	758	741	676	667	808	815
10:00	709	678	678	717	721	960	924
11:00	600	608	580	625	668	953	956

Burglaries

Hour	Sun	Mon	Tue	Wed	Thu	Fri	Sat
12:00 a.m.	45	40	37	38	37	38	57
1:00	41	26	32	32	32	27	34
2:00	30	27	34	28	33	28	34
3:00	26	29	26	28	22	31	31
4:00	33	31	31	30	28	43	36
5:00	13	17	19	16	15	19	14
6:00	18	50	47	35	47	49	28
7:00	35	179	172	156	187	147	44
8:00	10	105	91	91	88	96	16
9:00	35	109	119	111	113	113	45
10:00	39	78	83	91	72	79	44
11:00	35	88	61	69	81	79	53
12:00 p.m.	48	80	77	85	68	79	63
1:00	48	56	56	52	57	50	47
2:00	52	70	47	66	54	84	58
3:00	52	64	56	60	58	85	82
4:00	40	57	57	63	55	84	75
5:00	46	82	75	62	60	126	62
6:00	48	51	52	67	77	103	72
7:00	38	61	59	64	57	74	53
8:00	27	54	43	47	46	55	49
9:00	41	52	54	44	41	52	44
10:00	37	47	52	40	33	44	53
11:00	42	50	35	52	43	58	43

Robberies

Hour	Sun	Mon	Tue	Wed	Thu	Fri	Sat
12:00 a.m.	133	93	86	56	83	88	169
1:00	132	66	78	55	59	82	162
2:00	144	42	50	39	55	45	141
3:00	159	21	34	25	27	49	155
4:00	86	27	26	30	29	24	89
5:00	29	23	17	19	15	22	26
6:00	23	25	27	17	29	21	32
7:00	22	28	28	24	19	27	35
8:00	23	22	15	27	29	17	21
9:00	12	37	31	30	19	31	25
10:00	29	36	40	25	51	27	28
11:00	35	44	33	43	46	55	36
12:00 p.m.	52	47	47	46	41	46	42
1:00	48	57	56	50	55	52	65
2:00	48	54	53	64	39	56	61
3:00	58	68	78	74	83	80	64
4:00	45	81	83	90	62	79	44
5:00	58	96	56	78	76	74	55
6:00	59	98	69	80	68	87	66
7:00	80	95	64	91	72	92	83
8:00	98	112	97	105	81	116	84
9:00	120	140	127	120	96	146	125
10:00	113	118	124	153	128	144	147
11:00	107	118	102	113	108	148	137

Figure 2.10
Crime Hot Spots in Richmond, Virginia, 8:00 p.m.–12:00 a.m.

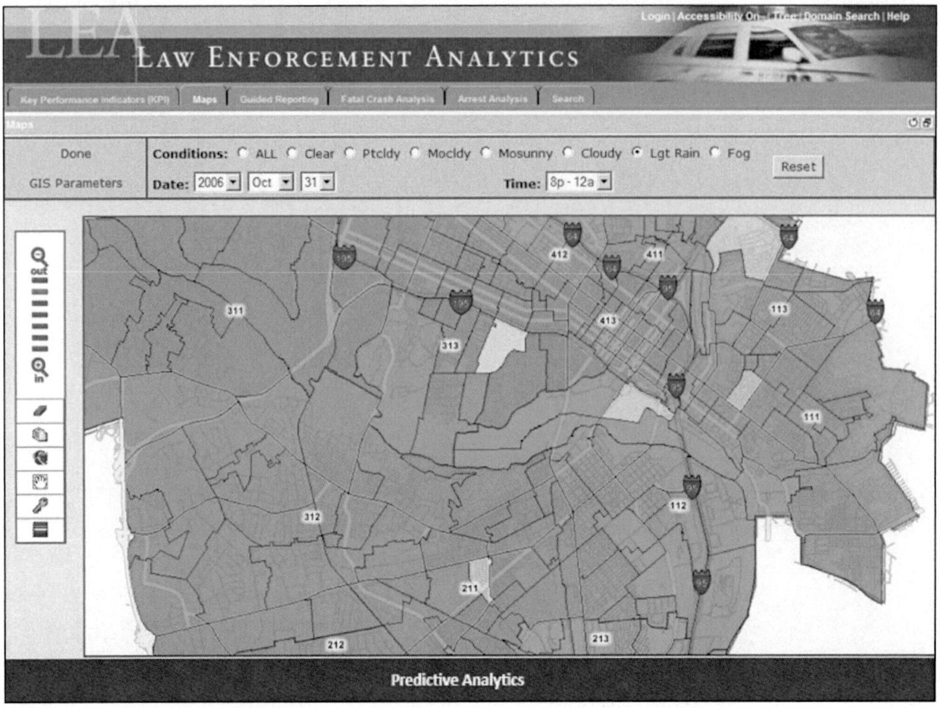

SOURCE: Information Builders, Law Enforcement Analytics interface. Used with permission.
RAND RR233-2.10

Spatiotemporal Modeling Using the Generalized Additive Model

A more complex method involves the use of a spatiotemporal generalized additive model (ST-GAM) and a local spatiotemporal generalized additive model (LST-GAM) developed by Xiaofeng Wang and Donald Brown at the University of Virginia, Charlottesville.[48] These models are extensions of regression models on grids; the input data include probabilities that each grid cell had a particular spatiotemporal feature at a particular time. Here, a "spatiotemporal feature" can be a prior crime, general attribute (e.g., socioeconomic indicator), or the presence of a geospatial feature (major infrastructure type) within a grid cell, all of which can be indexed by time (e.g., how long ago the last crime in a grid cell occurred).

Both models combine the spatiotemporal features of the crime area with crime incident data to predict the location and time of future crimes. ST-GAM is designed to predict crime for an entire region of interest. The assumption is that the underlying patterns are the same throughout a region. LST-GAM allows for differing regional

---

[48] Xiaofeng Wang, and Donald E. Brown, "The Spatio-Temporal Modeling for Criminal Incidents," *Security Informatics*, Vol. 1, No. 2, February 2012.

patterns. Both models produce a probability that a crime will be committed at a certain place and time conditioned on the spatiotemporal features of the area in which past crimes were committed.

Wang and Brown tested the two models against a spatial generalized linear model (GLM) and a hot spot method using time-indexed burglary data from Charlottesville, Virginia. Input data used in the ST-GAM and LST-GAM models included GIS data and demographic data by census block.[49] The ST-GAM and LST-GAM models outperformed the GLM model (which did not include temporal characteristics), and both significantly outperformed the hot spot method.

Seasonality

Some types of crime are strongly influenced by cyclical patterns, such as the day of the week or even the season. For example, during the summer, when children are not in school, there may be a spike in petty crimes and burglary. Including these cyclical effects in regressions is important for eliminating known sources of variation and will lead to better predictions. Cyclical trends need at least three data points per cycle. (Two points form a line, but more are required to provide some notion of the goodness of fit of the seasonality.) More data are generally better when establishing these effects in crime, though going back too far may diminish emerging trends. So, for seasonality, five years of data is a good target. For day of week effects, it may be acceptable to go back more than five weeks, unless there has been a major change in that period.

Adjusting trends for seasonal effects involves decomposing the crime data into systematic patterns. For example, if we were to examine five years of crime data by quarter, we might start by calculating the average crime incident rate for the entire five-year period. This gives us the overall *trend*, the first pattern. The next pattern is the residual: the difference between the trend just calculated and the reported crime incident for each quarterly entry. The important seasonal factor is calculated by finding the quarterly average across all five years of data. Finally, the seasonally adjusted pattern is calculated by subtracting the quarterly seasonal factor from the original series for the corresponding quarter. This is generally referred to as *additive decomposition*, and it assumes that the data are independent. Figure 2.11 is a notional quarterly record of crime incidents from a fictional city over the past three years. The trend in this case is flat. The unadjusted data show sharp increases in crime in the first quarter of each year and a dip in the last quarter. When seasonally adjusted, it is easy to see that the trend is indeed flat.

Other available methods include the U.S. Census Bureau's X-12-ARIMA (autoregressive integrated moving average), a widely used seasonal adjustment software with

---

[49] For a mathematical discussion of spatial GLM, see C. A. Gotway, and W. W. Stroup, "A Generalized Linear Model Approach to Spatial Data Analysis and Prediction," *Journal of Agricultural, Biological, and Environmental Statistics*, Vol. 2, No. 2, June 1997.

Figure 2.11
Trends in Crime Incidents, 2009–2011

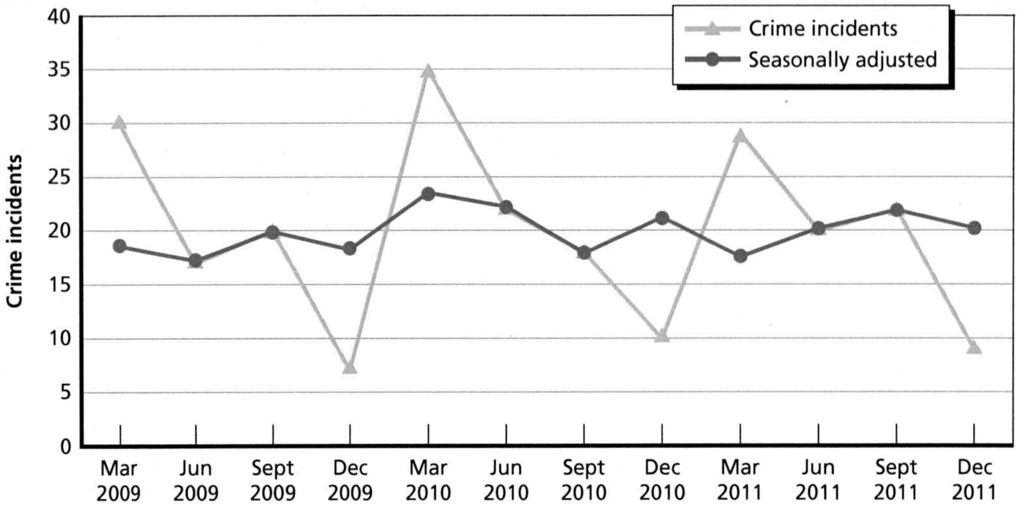

NOTE: Data in the figure are notional.

many seasonal and trend filter options,[50] and TRAMO/SEATS, a seasonal adjustment program developed by the Bank of Spain. TRAMO (Time-Series Regression with ARIMA Noise, Missing Observations, and Outliers) and SEATS (Signal Extraction in ARIMA Time Series) are linked programs. TRAMO provides automatic ARIMA modeling, while SEATS computes the components for seasonal adjustment.[51]

Risk Terrain Analysis

Risk terrain analysis comprises a family of techniques that (1) attempt to identify geographic features that contribute to crime risk (e.g., bars, liquor stores, certain types of major roads) and (2) make predictions about crime risk based on how close given locations are to these risk-inducing features.[52] In this section, we consider two major examples of risk terrain analysis: a simple and easy-to-employ heuristic method and a more complicated statistical modeling approach.

From the perspective of a police officer, the output of the risk terrain model will be qualitatively the same as that of a hot spot method: Both highlight areas that are

---

[50] U.S. Census Bureau, "The X-12-ARIMA Seasonal Adjustment Program," web page, undated.

[51] Catherine Hood Consulting, "TRAMO/SEATS FAQ," web page, last updated April 30, 2013.

[52] Features that tend to reduce risk are allowed in the models as well. There are a variety of names for this family of methods, including "geospatial predictive modeling," which we find to be overly general. *Risk terrain analysis* is a bit more specific and descriptive of what these methods do.

likely subject to high crime in the near future. However, from an analyst's perspective, they are very different methods. Hot spot methods are fundamentally clustering techniques that flag areas where clusters of crimes have occurred. Risk terrain modeling is a classification approach that characterizes a region's risk for crime based on its geographic traits.

A Heuristic Approach: Risk Terrain Modeling

Risk terrain modeling (RTM) is a simple approach to assessing how geospatial factors contribute to crime risk that was developed by Joel Caplan and his associates at Rutgers University. In their compendium on RTM, Caplan and Kennedy describe RTM as follows:

> Risk Terrain Modeling (RTM) is an approach to risk assessment in which separate map layers representing the spatial influence and intensity of a crime risk factor is created in a geographic information system (GIS). . . . [A]ll map layers are combined to produce a composite risk terrain map with values that account for all risk factors at every place throughout the landscape.[53]

RTM is a toolkit that plugs into ArcGIS. The method used is fairly simple. First, the analyst lays a grid over the jurisdiction to be analyzed. The analyst then tests the statistical relationship between the presence of certain geospatial features in grid cells (where the geospatial features are marked in GIS layers) and the presence of crimes of interest within that grid cell. Features with a strong positive association with crime are selected for the model. The method then counts the number of selected features present in each grid cell; grid cells with the greatest number of risk-inducing features are labeled as likely hot spots (and colored red or orange, typically).

Figure 2.12 shows an RTM output for a Shreveport, Louisiana, police district. Geospatial features include the presence of individuals on probation and parole, whether there was a crime in the previous six months, whether there was a crime the previous 14 days, whether there were calls for disorderly conduct or acts of vandalism in the previous six months, and whether there were buildings known to be "at risk." As shown, the more factors a grid cell contains, the "redder" it appears on the map.

In a 2011 paper, Caplan, Kennedy, and Miller used geospatial analysis to locate shooting hot spots in Irvington, New Jersey.[54] They built a data set of coordinates for known gang member residences, certain retail businesses (including bars, strip clubs, and liquor stores), and drug arrests. Applying these three sets of coordinates in a grid, they determined which grid cells were most likely to have shootings based on a six-

---

[53] Joel M. Caplan, and Leslie W. Kennedy, eds., *Risk Terrain Modeling Compendium for Crime Analysis*, Rutgers Center on Public Security, Newark, N.J.: Rutgers Center on Public Security, 2011.

[54] Joel M. Caplan, Leslie W. Kennedy, and Joel Miller, "Risk Terrain Modeling: Brokering Criminological Theory and GIS Methods for Crime Forecasting," *Justice Quarterly*, Vol. 28, No. 2, April 2011.

Figure 2.12
RTM Output for Shreveport, Louisiana

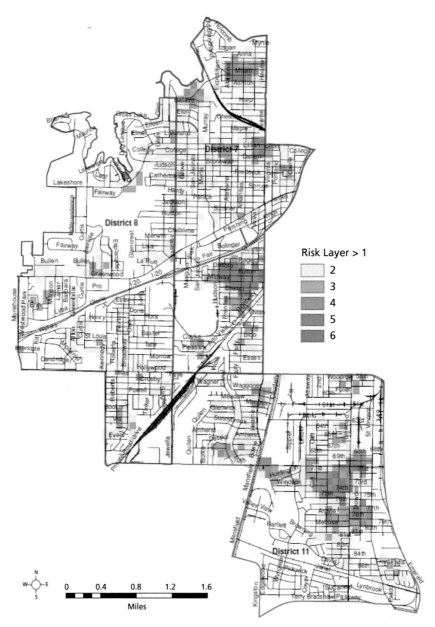

SOURCE: Susan Reno, Police System Administrator, Shreveport Police Department,
"PILOT: Predictive Intelligence Led Operational Targeting," presentation at the
National Institute of Justice Conference, Arlington, Va., June 19, 2012. Courtesy of the
Shreveport Police Department.

RAND RR233-2.12

month data set. This model was then applied to a different six-month period for validation. They found that the GIS approach was superior to simple retrospective modeling. (Retrospective modeling assumes that the locations of crimes in the prior six months are likely to be the locations of crimes in the next six months.)

A Statistical Approach to Risk Terrain Analysis

The statistical approach to risk terrain analysis involves two major phases. In the first phase, the algorithm compares the distances between crimes and types of geospatial features of interest (again, typically suspected risk factors, such as bars and liquor stores) and tracks the distances between the crimes and the nearest geospatial feature of each type. In the second phase, the algorithm assesses how "similar" each point on a grid is to locations that have seen crimes with respect to distances to the geospatial features. Points whose distances to geospatial features resemble those of crime locations are judged to be at higher risk. For illustration, suppose a large number of robberies occur about 50 meters away from a few bars in the city; grid points 50 meters from any bar in the city will tend to be assessed as being of high risk.[55]

DigitalGlobe's Signature Analyst is an example of a tool that employs this approach. Figure 2.13 shows example output from Signature Analyst, used here to predict purse snatchings in Washington, D.C. In the figure, the yellow-to-red colors show areas considered to be statistically similar to purse snatching locations (black dots) in June–July 2008. The analyst deliberately set the sensitivity of the algorithm to color enough areas to have a very high chance of capturing future purse snatchings. As shown, all purse snatchings in August–September 2008 ended up being in the areas colored on the map.

Discussion of Risk Terrain Analysis Approaches

There are two major advantages of risk terrain approaches. The first is that these methods are "genuinely" predictive in that they predict risk based on geographic attributes rather than merely extrapolate prior crime histories. Operationally, this means that these methods can predict new hot spots that are similar to other hot spots. The idea is that, even though the newly predicted hot spots have not seen recent crimes, they are similar enough to prior hot spots that they should be considered to be high-risk. The vendor behind Signature Analyst, for example, presents cases in which a number

---

[55] Brown, Dalton, and Hoyle explain the basic mathematics of the method: It uses a kernel density–like approach in which—rather than assess points' risk based on their proximity to recent crimes—the equations are modified to assess risk (i.e., applying the kernel function), the differences between the crimes' distances to geospatial features, and the grid points' distances to geospatial locations. See Donald Brown, Jason Dalton, and Heidi Hoyle, "Spatial Forecast Methods of Terrorist Events in Urban Environments," *Lecture Notes in Computer Science*, Vol. 3073, 2004.

Figure 2.13
Using a Risk Terrain Analysis Tool to Predict Purse Snatching Risk

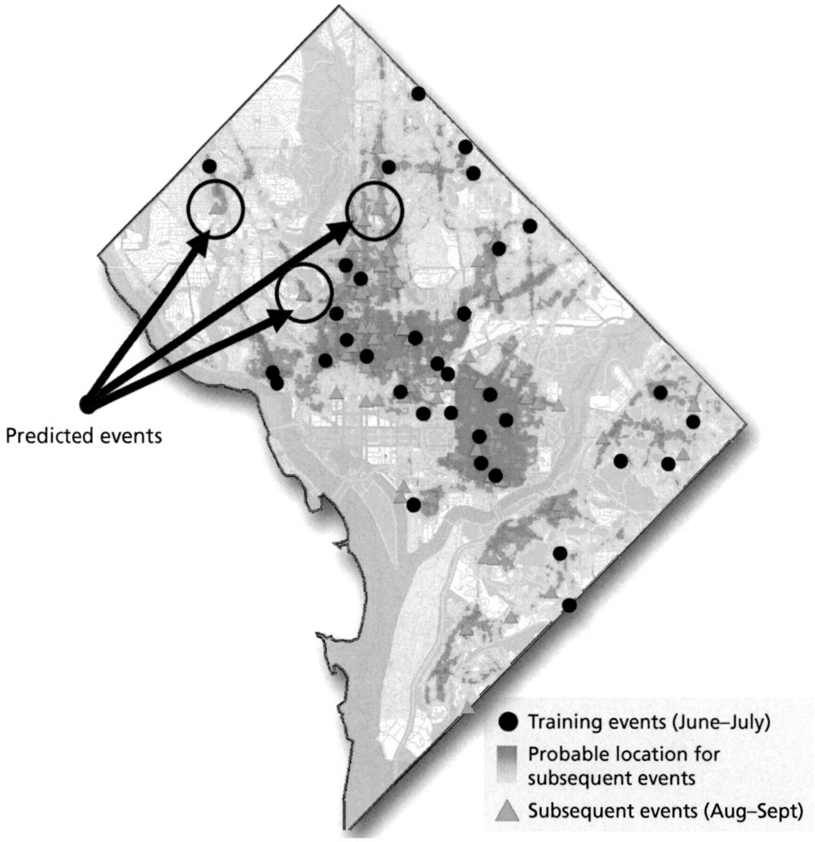

Predicted events

● Training events (June–July)
■ Probable location for
   subsequent events
▲ Subsequent events (Aug–Sept)

SOURCE: Peter Borissow, "Crime Forecast of Washington DC," Wikimedia Commons public
domain image, March 30, 2009.
NOTE: The figure is based on data on purse snatchings in Washington, D.C., in summer 2008.
RAND RR233-2.13

of crimes occurred in newly predicted hot spots outside of historical crime locations.[56]
The second, related advantage is that the tools can show which types of geospatial features were used in the models to forecast crime. One can even make forecasts and identify risky features using very small amounts of incident data. Signature Analyst was

---

[56] See, for example, GeoEye Analytics and Alexandria Police Department, *Analysis Report: Elevating Insight for
Law Enforcement Using Geospatial Predictive Analytics*, Alexandria, Va., April 1, 2010. The report describes the use
of Signature Analyst to project hot spots for car parts thefts in Alexandria, Virginia. The tool generated a number
of hot spots outside locations of recent thefts, and, indeed, future thefts occurred in the new hot spots that had
not seen prior thefts.

used to project hot spots for the perpetrator who shot at Northern Virginia military landmarks in October 2010 using just four prior shooting locations.[57]

The following major cautions are directly related to the advantages just described. These tools are intended to add novel hot spots beyond where crimes recently occurred; combined with a desire to capture as many crimes as possible in hot spots, one can easily generate maps showing most of the populated part of a jurisdiction as a hot spot. For example, Figure 2.13 flags the most heavily populated areas of Washington, D.C. As for factors, the earlier caution about how the appearance of a predictive element in a model does not equal causation applies here, too. Note that many geospatial features are just proxies for populated areas. Similarly, while one can generate hot spot projections based on a handful events, whether the results can be trusted is a point of debate. As an example, robberies are more likely to occur in areas with high foot traffic, and foot traffic is likely to be higher around subway stops. If a risk terrain analysis links robberies to subway stops, it may be detecting only the linkage between robbery and foot traffic and not offering particularly deep insight. These issues are discussed in more detail in Chapter Five.

Prediction Methods

The prediction methods presented here are by no means an exhaustive set. Other methods exist, and still others will be developed in the future. However, the major categories covered in this chapter capture the wide range of existing methods. Hot spot methods are used by most police departments, even though these methods lack the sophistication of others described in this chapter. Regression is a familiar tool used extensively by departments where sufficient data are available to produce statistical significance. The same is true of data mining methods: Where data are sufficient and data mining software is available, data mining can reveal interesting crime patterns. Near-repeat methods are based on the simple proposition that future crimes are likely to occur near current crimes, and geospatial methods highlight crime spots graphically.

It is important to bear in mind that the predictive methods discussed here do not predict where and when the next crime will be committed. Rather, they predict the relative level of risk that a crime will be associated with a particular time and place. The assumption is always that the past is prologue; predictions are made based on the analysis of past data. If the criminal adapts quickly to police interventions, then only data from the recent past will be useful to police departments.

The next chapter focuses on the last two steps in the prediction-led policing business process: police intervention based on predictions and criminal response.

---

[57] Colleen McCue, Lehew Miller, and Steve Lambert, "The Northern Virginia Military Shooting Series: Operational Validation of Geospatial Predictive Analytics," *The Police Chief*, Vol. 80, No. 2, February 2013.

CHAPTER THREE

Using Predictions to Support Police Operations

It's tough to make predictions, especially about the future.

—Yogi Berra

No matter what type of intervention is chosen, there is a need to provide tailored information to law enforcement personnel at all levels on both the forecasts and the data supporting the forecasts. Examples of the latter might include recent crime locations and descriptions, major call locations and descriptions, locations of crime attractors, reports on persons of interest in an area, and recent field interview reports. Officers need this information to respond to problems.

The primary focus of this chapter is the last two steps of the prediction-led policing business process depicted in Figure 1.1 in Chapter One. In Figure 1.2, we provided an overview of the basic intervention types: generic, crime-specific, and problem-specific. In this chapter, we provide examples of each. We begin, however, with a discussion of evidence-based interventions, providing a basis for engaging in proactive interventions.

Evidence-Based Policing

The Center for Evidence-Based Crime Policy at George Mason University conducted a study of 89 "rigorous policing intervention evaluations."[1] The center, led by David Weisburd and Cynthia Lum, asserts that decisionmakers can develop effective intervention strategies only by evaluating scientifically valid interventions.

After examining the 89 policing intervention evaluations, the center developed a three-dimensional evidenced-based policing matrix (depicted in Figure 3.1). The matrix is interactive and can be downloaded from the center's website.[2] The x-axis

---

[1]   Cody W. Telep, "Police Interventions to Reduce Violent Crime: A Review of Rigorous Research," Fairfax, Va.: Center for Evidence-Based Crime Policy, George Mason University, 2009.

[2]   Center for Evidence-Based Crime Policy, George Mason University, "Evidence-Based Policing Matrix," web page, undated. For additional background on the matrix, see Cynthia Lum, Christopher S. Koper, and Cody W. Telep, "The Evidence-Based Policing Matrix," *Journal of Experimental Criminology*, Vol. 7, No. 1, March 2011.

records the intervention target or scope of the target, the y-axis records the specificity of the intervention (similar to the crime-specific category), and the z-axis records the degree of proactivity in the intervention. The center rated the interventions as successful (black dot), having mixed results (gray dot), having insignificant results (white dot), or having harmful results (red triangle). As an illustration of how to interpret the figure, we note that it shows that 39 of the 89 cases examined were classified as individual interventions. The center reported 11 successful, eight mixed, 16 insignificant, and four harmful individual interventions. The center concluded,

> Overall, police can be most effective in reducing violent crime when they are proactive, use specific (as opposed to general) strategies, focus on small places (or groups operating in small places), and develop tailor-made solutions that make use of a careful analysis of local problems and conditions.[3]

Figure 3.1
Evidence-Based Policing Matrix

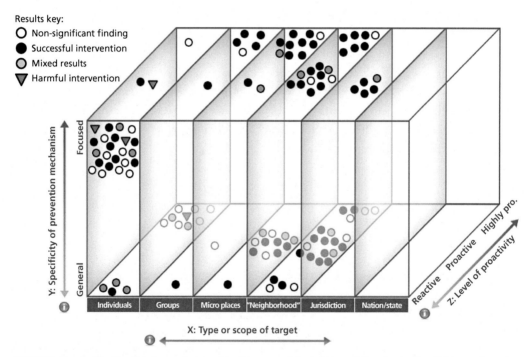

SOURCE: Created by Cynthia Lum, Christopher Koper, and Cody Telep at the Center for Evidence-Based Crime Policy, George Mason University, undated. Used with permission. The interactive version of the matrix can be found at http://www.policingmatrix.org.
RAND RR233-3.1

---

[3] Telep, 2009.

Taking Action on Hot Spots in Washington, D.C.

As a hypothetical example, consider how an analyst might approach a problem of robberies in Washington, D.C. Figure 2.7 in Chapter Two presented a regression model that showed forecasts of robberies, with one of the hot spots labeled simply "Columbia Heights." An entire neighborhood does not make an operationally actionable hot spot.

Figure 3.2, on the other hand, shows individual robbery locations. The figure is a screenshot from CrimeReports.com (taken on December 10, 2012); it shows robberies in the area (geolocated to the block level for privacy reasons) between June 10 and September 10, 2012.[4] The purple icons containing the letter "R" represent individual robberies; the light blue icons represent multiple robberies on the same block. While there is a good bit of distribution of robberies, we can see that the multievent icons are concentrated in the streets directly around, and leading out of, the Columbia Heights Metro station and the DC USA shopping mall that abuts the Metro station. This concentrated area, accounting for almost 40 robberies alone, is denoted by the overlaid

Figure 3.2
Robberies in the Vicinity of the Columbia Heights Neighborhood, Washington, D.C.

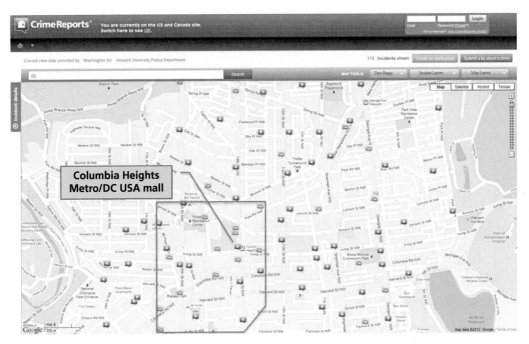

SOURCE: CrimeReports.com, generated December 2012. Used with permission.
RAND RR233-3.2

---

[4]   CrimeReports.com is a PublicEngines product. PublicEngines also offers CommandCentral Analytics and CommandCentral Predictive. The latter provides law enforcement agencies with weighted recommendations about where to direct patrols.

boundaries in the figure. In this case, we might develop strategies specifically designed to protect Metro commuters and pedestrians visiting the mall.

At the problem-specific level, the next step would be to drill down on the individual locations (addresses or block faces) that have seen multiple robberies, review the individual crime reports and determine specific risk factors, and identify appropriate problem-solving strategies. For example, we might focus on the blocks in the hot spots that saw multiple robberies. These would include the four street blocks immediately adjacent to the Metro station, which collectively saw 11 robberies over the period studied. Also of interest are Columbia Road NW and several adjacent streets leading away from the Metro station to the southwest, as well as the stretch of Park Road NW leading away from the northeast, all of which saw multiple robberies over the same period and account for 16 robberies in total. (The stretch of Columbia Road between 14th Street NW and 15th Street NW accounted for five robberies alone.) Figure 3.3 highlights these areas.

Figure 3.3
Robbery Locations Near the Columbia Heights Neighborhood, Washington, D.C.

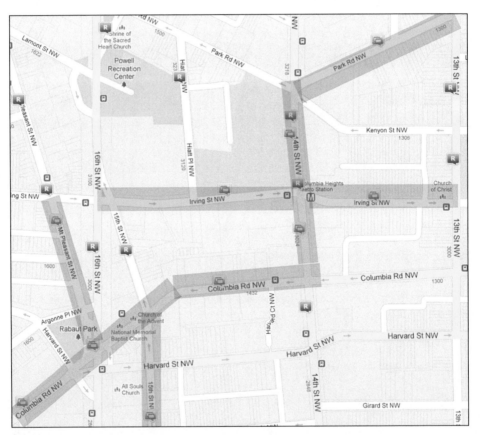

SOURCE: CrimeReports.com, generated December 2012. Used with permission.
RAND RR233-3.3

Once individual incidents in the hot spots are identified and characterized, the analyst can begin looking for attributes common to the bulk of the crimes. Examples might include specific addresses, specific types of locations, specific times of day, and specific attributes of the victims being targeted. The Washington, D.C., example includes data that are no more detailed than the block level and date. Even from there, one can start considering specific interventions to reduce crime. One such intervention in Washington, D.C., involved the Metropolitan Police issuing cards to Metro riders using the Columbia Heights station (and other Metro stations with a higher-than-average numbers of robberies) warning of the dangers present at the station and the surrounding area and how to avoid being robbed.[5]

Koper Curve Application in Sacramento

A good example of a generic intervention is the Sacramento Police Department's test of Koper's 13- to 15-minute stay rule. This example also shows how academic research can inform field operations to help develop best practices.

In February 2012, the Sacramento Police Department initiated a study to test Koper's finding that the optimal patrol duration to deter crime in hot spots is between 13 and 15 minutes.[6] The focus was on Part 1 crimes (e.g., homicides, aggravated assaults) and calls for service. The department identified a total of 42 hot spots across the city. It then created a treatment group and a control group, each consisting of 21 hot spots. To ensure statistical significance, hot spots in both groups were paired so that each group consisted of a number of high-ranking and low-ranking hot spots. Figure 3.4 shows the 42 hot spots color-coded by group, treatment and nontreatment. The study was carried out between February 8 and May 8, 2012. The objective was to determine the effect on Part 1 crimes in the treated and nontreated hot spots during the same time period.

The patrols in the treated hot spot districts were told to be proactive: They were to randomly visit each hot spot for 12–16 minutes and to revisit every two hours. This latter requirement stemmed from Koper's finding that reductions in crime last for about two hours after the previous visit.

The results of the test were rather dramatic. Part 1 crime rates during the 90-day test were compared with the same crime rates in the same period in 2011. In the treatment districts, Part 1 crimes decreased by 25 percent, whereas in the nontreated districts, Part 1 crime rates actually increased by 27.3 percent. The same comparison was

---

[5] The analysis presented in this section is strictly a hypothetical example devised for this report. It has no connection to do with the Washington, D.C., Metropolitan Police Department's decision to issue robbery-prevention cards.

[6] Danielle Ouelette, "A Hot Spots Experiment: Sacramento Police Department," *Community Policing Dispatch*, Vol. 5, No. 6, June 2012.

Figure 3.4
Hot Spot Locations in Sacramento, California, February 8–May 8, 2011

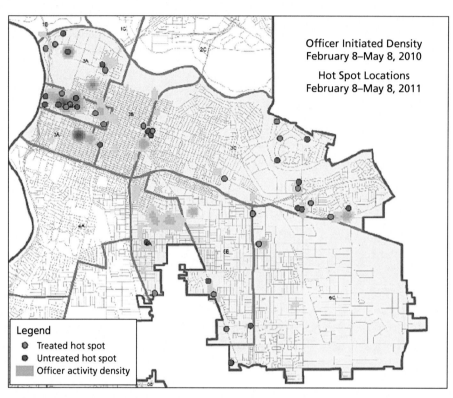

Officer Initiated Density
February 8–May 8, 2010

Hot Spot Locations
February 8–May 8, 2011

Legend
- Treated hot spot
- Untreated hot spot
- Officer activity density

SOURCE: Ouelette, 2012, Figure 1.
RAND RR233-3.4

made for calls for service. In the treated districts, they decreased by 7.7 percent, and in the nontreated districts, they increased by 10.9 percent.

Numbers like these certainly call into question issues of displacement. Were criminals simply choosing to commit crimes in districts where the patrols were not as proactive? The department reportedly *did* examine this phenomenon. It examined Part 1 crimes in the two-block radius surrounding each hot spot and found that there was no significant increase in crimes in these "buffer zones."

Investigating Convenience Store Robberies in Chula Vista, California

The city of Chula Vista, California, experienced a total of 157 store robberies in the 45-month period beginning from August 2002 through April 2006. Perpetrators targeted 7-Elevens, gas station/convenience stores, and liquor stores/mini-marts. Figure 3.5 maps these robberies and also shows the 29 stores that were never robbed.[7]

From the hot spot analysis alone, it appears that the entire east end of the city consists of hot spots. Clearly, it is difficult to take effective action, given an area this large and limited policing resources. Further analysis of the robbery data revealed that 19 stores accounted for 110 of the robberies, an illustration of the "80/20" rule. Upon further examining the data, analysts discovered that 7-Elevens appeared to be the most attractive to robbers. Eight of the 12 stores robbed five or more times were 7-Elevens. Only one of the 14 7-Elevens in the city was not robbed at all during this period.

In this way, the crime pattern was narrowed to robberies of 7-Eleven stores in the city, and the hot spot was reduced from the entire east end to 14 stores. Crime-specific intervention strategies could then be focused on understanding why these stores were robbed repeatedly.

Figure 3.5
Convenience Store Robberies in Chula Vista, California, 2002–2006

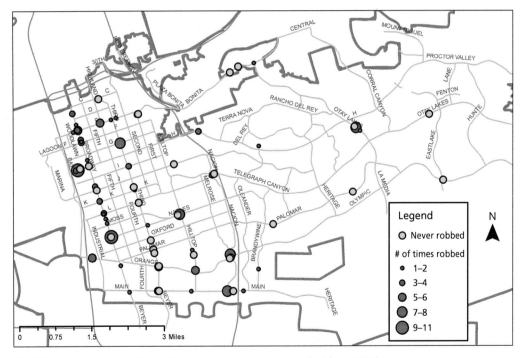

SOURCE: Courtesy of the Chula Vista Police Department. Used with permission.
RAND *RR233-3.5*

---

7 Julie Wartell, Independent Adviser on Public Safety, "GIS for Proactive Policing and Crime Analysis," presentation at the Technologies for Critical Infrastructure Protection Conference, National Harbor, Md., August 31, 2011.

Upon inspection, analysts realized that although the robbery "hot spots" were spread fairly widely geographically, from an attribute perspective, the robberies mostly targeted convenience stores from one particular chain. The local police department was then able to work with the company and store proprietors to improve safety measures.

Predictive Policing in Context: Case Studies

In this section, we present examples of how police departments have applied the techniques discussed in Chapter Two. In most cases, more than one method was employed, along with other information-led policing approaches. The objective of this section is to illustrate how predictive methods can be used in conjunction with police operations in general. In most of the cases here, operations follow from the prediction step in Figure 1.1 in Chapter One to police operations and criminal response. The emphasis in these cases is on illustrating the predictive method used, the police response to the prediction (if any), and the outcome (if that information is available).

Most of these case descriptions were provided by the various departments that implemented the methods. In those cases, there has been no independent verification of the claims concerning the success of these methods. We include them here to illustrate how various departments are utilizing predictive and analytic methods to fight crime in their communities and how these methods are integrated into information-led policing methods. It should also be noted that the predictive software and methods used in these case studies were largely basic statistical tools that could be suitable for a large fraction of departments.

Shreveport, Louisiana: Predictive Intelligence–Led Operational Targeting

The Shreveport Police Department is conducting an NIJ-funded experiment originally focused on developing a leading indicators model to forecast tactical crime numbers at the reporting-area level (similar to a precinct). The objective of the experiment, Predictive Intelligence–Led Operational Targeting (PILOT), is to flag potential crime spikes one month ahead, given recent input data values related to crime and disorder. Shreveport defined "tactical crime" as robbery, burglaries, vehicle break-ins, outside thefts (e.g., thefts other than shoplifting), and stolen vehicles. Input variables to be tested included various types of calls for service, lesser crimes, juvenile arrests, and seasonal indicators. All values were lagged one month except for seven-day prior and 14-day prior values for tactical crime. The experiment also included "spatial-lag" variables for the calls and lesser crimes—incident counts within adjacent districts. The approach followed the work of Gorr and Olligschlaeger, who developed similar models at the precinct and large grid (4,000 sq. ft.) levels.[8]

---

[8]   Gorr and Olligschlaeger, 2002.

In the experimental districts, Shreveport's prediction evolved from a district-level crime forecasting model to a hot spot forecasting model, identifying areas at risk on a much smaller scale. Shreveport analysts overlaid a grid of 400 sq. ft. cells on the experimental area and used two predictive methods to identify hot spots:

- First, analysts used logistic regression to forecast the likelihood of a tactical crime in each cell using a subset of variables similar to those mentioned earlier. (They used stepwise regression to select the variables.)
- Second, they used RTM to identify "higher-risk" cells. Analysts identified a set of geospatial features or attributes associated with a higher risk of crime (e.g., having greater numbers of prisoners or parolees or previous high crime or disorder counts). Individual cells then received one point for each feature or attribute, with cells having more points assessed as higher-risk.

For both methods, Shreveport analysts provided units in the experimental districts with ArcGIS-generated maps like as those depicted in Figure 3.6. In the figure, cells at higher risk are coded with shades ranging from yellow (elevated) to red (highest risk). The RTM map in the left of the figure and the logistic regression map to the right highlighted the same general areas as being at higher risk. The RTM map tended to produce larger, fewer hot spots, and the logistic regression map tended to produce larger numbers of one- or two-cell hot spots.

Once the experiment was implemented, Shreveport analysts used the logistic regression predictions exclusively, as practitioners preferred to focus on smaller spots. Predictive maps were updated monthly.

Staff in Districts 7 and 8 responded to the predictions by dispatching dedicated teams (two two-person cars plus a sergeant) for focused patrols in these hot spots. The teams were officially on overtime and were dedicated to policing the predicted hot spots. The teams focused both on looking for suspicious activity in their hot spots and on building positive (and intelligence-generating) relationships with the local community. For example, officers looked for people walking down the middle of the street when sidewalks were available and stopped and questioned them; those with criminal records for property crime were cited for violating a city ordinance and questioned aggressively. Officers also followed up with neighbors of burglarized properties, letting them know what had happened and asking whether they had any information that could help solve the crime. Officers generally made many contacts with members of the community to let them know what they were doing and to ask whether they could provide crime-fighting information. Such measures reportedly led to much higher volumes of tips and much greater support from the community.

Shreveport Police Department staff, with the assistance of crime analysts, collected field interview and tip information and distributed that information back to officers so they could take action the next day. The crime analysts worked with district

Figure 3.6
At-Risk Districts in Shreveport, Louisiana

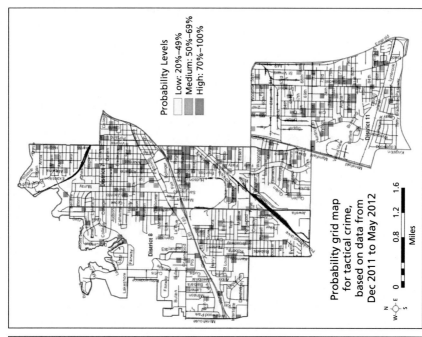

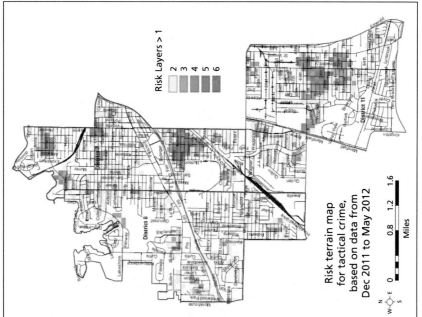

SOURCE: Courtesy of the Shreveport Police Department.
RAND RR233-3.6

officers to develop daily maps showing not just the predictions but also key situational awareness data, including information on recent crimes, suspicious activity, field interviews, and where predictive policing units had just visited. The analysts then worked with district commanders to select target areas for each day. When time-sensitive leads on crimes came in, other commanders and officers in the area quickly followed up on the leads, which generated a number of arrests. The districts placed great emphasis on "quality" contacts and arrests (related to Part 1 crimes), as opposed to raw numbers of stops and arrests. Results from the experiment were still being analyzed as of this writing. However, preliminary results suggest that major property crimes declined by more than 40 percent in the experimental hot spots over the controls while PILOT was running in Districts 7 and 8.

Memphis, Tennessee: Crime Reduction Utilizing Statistical History

Serious and violent crime was a problem in Memphis, Tennessee. In April 2009, Forbes Magazine ranked Memphis one of the most dangerous cities in the United States—second only to Detroit, Michigan. City officials countered by stressing that operation Blue CRUSH (Crime Reduction Utilizing Statistical History) had helped put crime on the decline since the program was first piloted in 2005.[9] Blue CRUSH is a data mining approach to the analysis of location- and time-based criminal patterns and evolving trends. According to Memphis police, because the program allows analysts to rapidly evaluate incoming patrol data against historical trends, they are able to respond to predicted threats before a criminal act is committed.

Although they do not deny the crime problem, local officials argue that Memphis suffers in national crime rankings because its police department tracks and captures the criminal environment with a high level of precision. Memphis was an early adopter of the FBI's National Incident-Based Reporting System, which is a refinement of the UCR system. As a result, the public has access to more data on crimes committed in the Memphis metropolitan area than they do for other regions. According to then-director of the Memphis Police Larry Godwin in a 2009 article, "You have got to measure yourself in order to improve." Godwin explained that crime in Memphis had declined by 16 percent between 2006 and 2008. In comparison to other major cities, such as Baltimore, which has a population of 634,549 and experienced 234 murders in 2008, Memphis experienced nearly 100 fewer murders out of a popu-

---

9   Zack O'Malley Greenburg, "America's Most Dangerous Cities," *Forbes Magazine*, April 23, 2009. Note that the article uses data from the Federal Bureau of Investigation's (FBI's) Uniform Crime Report (UCR) to "rank" the most violent cities based on violent crime per 100,000 people. The FBI does not advise this practice because such a basic summary of individual variables does not adequately capture the criminal environment in a particular town, city, county, state, or region.

lation that is approximately 37,000 larger. By many accounts, Memphis is becoming a safer place.[10]

Memphis has used the detailed crime data collected by its police department to predict and prevent crime. Crime maps and data analysis are now integrated directly with police planning around problem neighborhoods, the allocation of patrol and special units, and strategic planning for citywide crime reduction. Blue CRUSH was established through a partnership between the Memphis Police Department and the University of Memphis to counteract a sharp spike in gang-related gun violence, using IBM's SPSS. Speaking about the program, John Williams of the Memphis Police Department said, "criminals leave a footprint," adding that the SPSS software licensed by the Blue CRUSH program had helped police use those tracks to predict future hot spots.[11] Memphis had seen a rise in gang-related activity in particular areas, but by crunching existing and even incoming data from patrol units, Blue CRUSH pointed the police toward specific locations and times. Knowing the street corners and city blocks that needed the most attention provided the department with an opportunity to plan tactics and take action in specific locations. The actions taken depended on a range of factors related to the location, the type of predicted crime, and other available information, but they included the deployment of marked and unmarked patrols, increased numbers of vehicle stops, and assigning undercover units to infiltrate locations of interest. The pilot program demonstrated a considerable improvement over the department's standard saturation and zero-tolerance tactics. As a result of the successful pilot, Blue CRUSH was expanded citywide in early 2007.[12]

The software used by Blue CRUSH relies on existing criminal records and incoming patrol data, including crime locations, crime types, time of day and day of week, and a variety of victim characteristics, to generate tactical crime predictions. The software arrays the historical and incoming data on a multilayered map of the region, allowing police to evaluate all precincts in the metropolitan area at once or by specific blocks. Because Blue CRUSH captures data points throughout each day and connects them to past events, predictions can be timely. In these predictions, Memphis police can respond by strategically placing unmarked cars to catch criminals in the act or by deploying Blue CRUSH–marked vans to specific blocks to reduce the opportunity for criminal activity. Richard Janikowski, an associate professor of criminal justice at the

---

[10] Chris Conley, "Memphis a Victim of Crime Reports," *The Commercial Appeal* (Memphis, Tenn.), June 29, 2009.

[11] Conley, 2009.

[12] Candy Phelps, "IBM Predictive Analytics Help Slash Crime Rates in Memphis," *Public Safety IT Magazine*, November 2010.

University of Memphis, has described the objective of Blue CRUSH as "getting the right police resources in the right place, on the right day, at the right time."[13]

In 2005, before Memphis initiated the pilot, the FBI reported that crime rates in the city had increased by 2.5 percent since 2004, above the national average of 2.3 percent for the same period.[14] As of October 2010, carjacking was down by 75 percent, and business robberies were down by 67 percent citywide (compared with 2006 data). In 2007, Blue CRUSH investigations of drug sales led Memphis police to four metropolitan-area motels. Informed by Blue CRUSH data, the department tasked undercover units to investigate the motels. The investigation revealed drug sales, drug use, prostitution, underage prostitution, robberies, and a homicide. "Operation Heartbreak Hotel" helped the Memphis Police Department make the appropriate arrests and enabled the district attorney to shut down the mismanaged motels.[15] Since 2006, the department's efforts had helped reduce crime in the Memphis area by 31 percent and violent crime by 15.4 percent as of 2010.[16]

Blue CRUSH has been expanded to work in conjunction with the Memphis Police Department's Real Time Crime Center, a $3 million crime monitoring and analysis complex that was opened in 2008. The center was created to facilitate information sharing between various agencies in the Memphis metropolitan area. In 2010, Nucleus Research, Inc., led a cost-benefit study of the Blue CRUSH system. The study revealed an average annual benefit of $7,205,501 for an annual cost of $395,249 to the metropolitan area. As calculated by Nucleus Research, the benefit equates to the number of employed officers and police resources required to reduce crime at the levels observed since 2006. The IBM SPSS software is effectively paid for after 2.7 months of Blue CRUSH operation each year.[17]

Nashville, Tennessee: Integrating Crime and Traffic Crash Data

In the criminology sciences, the "deviant place" theory posited by Rodney Stark in 1987 suggests that there are places that attract people who commit crimes regularly.[18] As part of their normal activity, people who commit crimes will frequent

---

[13] Amy O. Williams, "Blue C.R.U.S.H. Walks Its Beat Among Community Organizations," *Daily News* (Memphis, Tenn.), November 16, 2006.

[14] According to the FBI's UCR, annual compilation of data from departments across the country. See Federal Bureau of Investigation, "Crime in the United States 2005," web page, September 2006.

[15] Shelby County District Attorney General, "Four Area Hotels Closed for Business Following 'Operation Heartbreak Hotel,'" news release, February 12, 2008.

[16] Jeffery Smith, "Memphis Police Leverage Analytics to Fight Crime," CivSource, July 21, 2010.

[17] Nucleus Research, *ROI Case Study: IBM SPSS—Memphis Police Department*, Boston, Mass., Document K31, June 2010.

[18] Rodney Stark, "Deviant Places: A Theory of the Ecology of Crime," *Criminology*, Vol. 25, No. 4, November 1987.

these areas, streets, and intersections. Ronald Wilson extended the deviant place theory in a 2010 article proposing that because the people who commit crimes exhibit behavioral problems—lack of self-control, lack of moral development, and persistent risk taking—the places that attract criminals are more likely to experience high rates of violations, specifically those leading to traffic collisions. The theory also posits that the traffic crime will attract other types of crime without persistent enforcement.[19] The theory has led to a series of police interventions known as data-driven approaches to crime and traffic safety (DDACTS); these programs have been adopted by police departments across the country.[20] The DDACTS model is supported by a partnership between the U.S. Department of Transportation's National Highway Traffic Safety Administration, the Bureau of Justice Assistance, and NIJ. These and other partnerships have helped secure a unique intersection of data allowing for hot spot techniques that police have used to identify areas with high concentrations of vehicle and crime incidents.

Nashville, Tennessee, experienced a spike in crashes related to drunk driving in the period leading up to 2003. In January 2004, the Metropolitan Nashville Police Department developed a plan to counteract this trend. The initiative, led by then–Chief Ronal Serpas, introduced an accountability-driven leadership model and a new approach to using data.[21] Chief Serpas planned to weave data and analysis into the regular department-level strategic planning process. He provided incentives to support the initiative by holding officers accountable for specific results. In Nashville, this type of crime analysis was used for planning, to demonstrate results to the public and city officials, and to develop metrics for accountability. Like other DDACTS programs across the country, Nashville uses high visibility enforcement strategies to reduce crime and traffic violations in predicted high-incident areas. By focusing primarily on traffic violations, these DDACTS programs have prompted area-wide crime reduction. Analysis of crash and crime data show that law enforcement agencies have affected the rates of both of these social harms by implementing highly visible traffic enforcement strategies informed by their DDACTS programs.[22]

Chief Serpas fostered accountability within the Metropolitan Nashville Police Department by holding department leaders responsible for setting weekly DDACTS

---

[19] Ronald E. Wilson, "Place as the Focal Point: Developing a Theory for the DDACTS Model," *Geography and Public Safety*, Vol. 2, No. 3, June 2010.

[20] DDACTS integrates location-based crime and traffic crash data with the goals of identifying the most effective allocation of law enforcement and other resources and reducing crime, traffic accidents, and moving violations. "[It] is an effective, predictive, location-based policing approach to crime and traffic safety that delivers law enforcement services at the right place and at the right time" (International Association of Directors of Law Enforcement Standards and Training, homepage, undated).

[21] Jason Wyatt, "Integrating Crime and Traffic Crash Data in Nashville," *Geography and Public Safety*, Vol. 2, No. 3, June 2010.

[22] Scott Silverii, "Changing the Culture and History of Policing," *Law Enforcement Today*, October 9, 2012.

goals, identifying hot spots, and deploying tactics during a series of comparative statistics (CompStat)–style meetings. CompStat is a statistics-driven management philosophy used by police departments to incorporate analysis (most commonly GIS and crime mapping) in identifying problem areas. These problem areas are then ranked, and responsibility for addressing them is assigned to commanders within the department. Chief Serpas held weekly meetings to discuss progress made in these problem areas, and data were analyzed regularly to track improvement or spot opportunities for new policing initiatives. In taking on more responsibility for operations in a given area of the metropolitan area—down to specific street corners—department leaders were newly accountable for modifying their actions and for decisions and strategies directed to solving specific community problems. Simultaneously, the Nashville police started to carefully collect and code large quantities of traffic, crime, and drunk-driving data across the city. With oversight by a team of specially trained analysts, the data were used to produce multilayered crime maps that arrayed traffic violations alongside other criminal activities. Regression analysis techniques were then used to determine the association between the layers. These data enabled the department's leadership to design (in its weekly meetings) community-specific tactics to reduce drunk driving.

The department increased its number of vehicle stops to strategically reduce traffic violations and other criminal activities. Areas that were selected for more vehicle stops had demonstrated high numbers of crime incidents, crashes, traffic violations, and drunk-driving arrests. Between January 1 and November 28, 2009, a total 271,994 vehicle stops resulted in 24,211 arrests for driving under the influence (DUI; 26.4 percent of all arrests during the period). Nashville's DDACTS initiative was deemed a success: Fatal crashes in the Nashville metropolitan area declined by 15.6 percent between 2003 and 2009, accidents that resulted in injuries were down by 30.8 percent, and DUI arrests increased by 72.3 percent. Analysis of the FBI UCR data for the Nashville metropolitan area also revealed that the rate of Part 1 crimes committed between 2003 and 2008 decreased by 13.9 percent.[23]

The Metropolitan Nashville Police Department has continued to build its analytic capabilities by investing in ArcGIS mapping software and data management systems. Although it was originally intended to reduce drunk driving, the Nashville's DDACTS initiative helped the region reduce traffic-related injuries and fatalities along with crime in general. Chief Serpas attributed much of this success to the program's high-quality data, good analysis, and a department-wide emphasis on fighting crime

---

[23] As mentioned earlier, the FBI UCR system collects data on serious crimes, such as murder and non-negligent manslaughter, forcible rape, robbery, aggravated assault, burglary, larceny, motor vehicle theft, and arson. These serious crimes are classified as Part 1 of the UCR program data. Part 2 data include information on arrests for other crimes, such as driving under the influence and simple assault. For more information on UCR and the Part 1 and 2 distinctions, see Federal Bureau of Investigation, "UCR General FAQs," undated.

rather than merely counting numbers.[24] By applying these lessons, the initiative was successful across the city, not just in known high-crime neighborhoods.

Baltimore, Maryland: Crash-Crime Project

Although crime had been decreasing in Baltimore, robbery, burglary, and auto theft remained a concern. In 2007, the Baltimore County Police Department identified a series of traffic corridors that were highly associated with criminal activities and car crashes. This finding led the department to join the national DDACTS predictive policing initiative. It hoped to decrease crime by addressing traffic violation hot spots across the more than 3,000 miles of roadway in Baltimore County. In 2008, the department launched the Crash-Crime Project to take action on car crash hot spots, minimize criminal activity displacement around the target areas, and reduce overall crime rates in the county.[25]

The project had two major phases. First, the Baltimore County police crime analysts studied traffic accident and criminal activity data to identify the most dangerous stretches of roadway. They used GIS mapping tools to create multilayered maps detailing the criminal environment and traffic violation and car crash patterns that defined "deviant places," neighborhoods, and street segments. They then used Nnh techniques to map both crime and traffic violations by day and time, identifying hot spots at the intersections of these clusters—employing the covering ellipses method. Data on hot spot locations and the days and times of the highest incident rates informed strategies to target vehicle stops and deploy high-visibility patrols to "calm" these areas. The department reported a 13.6-percent reduction in robberies, a 6-percent decrease in car crashes, and a 14.7-percent decrease in crashes resulting in injuries. Despite the success of the program's first phase, the department identified shifts in hot spots between 2007 and 2008 that suggested crime displacement, a problem addressed in the program's second phase.

To better direct police resources toward these potential crime displacement areas, Baltimore County crime analysts mapped instances of burglaries, robberies, and auto theft alongside the detailed crash and crime hot spots from the first phase of the project. For each overlapping area, they used a kernel density function to depict criminal concentrations within each hot spot. The police department, in turn, used these maps to develop neighborhood-, community-, and target-specific tactics to reduce concentrations of criminal activity through high-visibility traffic enforcements. The tactics included increased patrolling and more frequent vehicle stops or checkpoints in partic-

---

[24] Anacapa Sciences, Inc., *Data-Driven Approaches to Crime and Traffic Safety (DDACTS): Case Study of the Metropolitan Nashville, Tennessee, Police Department's DDACTS Program*, Santa Barbara, Calif., October 16, 2009.

[25] Howard Hall and Emily N. Puls, "Implementing DDACTS in Baltimore County: Using Geographic Incident Patterns to Deploy Enforcement," *Geography and Public Safety*, Vol. 2, No. 3, June 2010.

ular neighborhoods. Police were deployed to strategically address these new hot spots between April 1 and December 31, 2009.

To evaluate the department's progress, analysts compared average crime rates in the three years prior to the April 1–December 31, 2009, Crash-Crime Project intervention. They found that robberies had decreased by 33.5 percent, burglaries by 16.6 percent, and auto thefts by 40.9 percent. Total crashes decreased by 1.2 percent, and crashes involving injuries decreased by 0.2 percent. Overall, the project was considered successful, and the department has made efforts to improve its data collection and analysis capabilities as a result. By targeting specific areas during specific times with limited resources, the department was able to reduce crime across the county's large geographic area by developing and implementing data-driven interventions.[26]

Iraq: Locating IED Emplacement Locations

Immediately following major combat operations in Iraq near the end of April 2003, insurgent elements began emplacing IEDs along roadways frequented by U.S. and coalition forces. Predicting when and where these devices would be placed became increasingly urgent. However, the methods then in use to identify what were referred to as hot spot areas were not "actionable"—that is, they were too large and too irregular to help the commands make responsive tactical decisions, such as where best to position sniper teams, how to focus surveillance assets, or how to plan convoy or patrol routes to either suppress or avoid the emplaced IEDs. The RAND-developed Actionable HotSpot (AHS) concept grew out of these concerns. AHS was an attempt to use recent data on the time and locations of IED-related activities (e.g., detonated, found and cleared, interrupted during emplacement) to detect clustering patterns indicating possible future threat activities in the *immediate* area.

The AHS software looked for clusters of two or more IED events in small, user-specified areas (typically circles up to a 100-m radius) over a period of one to several weeks, with the time frame selected empirically to suit a particular area and intervention strategy. Analysts then scored and ranked the clusters using an exponential weight system that time-discounted earlier threat events, displaying the locations on a map. The weighting approach was tested during the calibration phase; the measure of effectiveness used to calibrate the code and evaluate the systems performance was the occurrence (or nonoccurrence) of future IED events in the AHS areas within a specified future time frame (typically, 24–48 hours after the AHS selections).

Analysts developed the AHS code in R (a programming language used in statistical applications) and used ArcGIS for the graphics displays and Microsoft Access to manage the historical data inputs. The AHS was tested on past data and demonstrated variable—but often encouraging—performance in several regions in Iraq. In

---

[26] Howard B. Hall, "Targeting Crash and Crime Hot Spots in Baltimore County," *The Police Chief*, Vol. 76, No. 8, July 2009.

addition to these tests against historical data, analysts used the experimental AHS code to provide AHS nominations to several brigade-size units on a daily basis from August 1 through December 23, 2006. The nominations were then sent to test units in the field for their consideration in tactical planning. The feedback from these units was positive: Most actions taken (sniper, overwatch) as a result of planning informed by AHS nominations led to positive results. Feedback from operators was used to further refine the code to improve its relevance for operational needs. Units in theater supplied analysts with updated IED event data daily, and the analysts supplied the units with timely AHS nominations (in the form of maps with ellipses identifying the nominated areas, alongside text data).

The results of the test period were encouraging. Although units did not always "action" the nominated hot spots, they were grateful for the increased situational awareness. When they did choose to take action in response to a nominated AHS, it was usually because the hot spot had been corroborated by other intelligence, and these actions were almost always successful. During the AHS test period, the units studied achieved an average success rate of 30 percent, with a range between 50 percent and 11 percent. By *success*, we mean that in the 24 or 48 hours following a nomination (the period varied by unit), at least one IED incident (explosion or found and cleared) occurred in the nominated area. The accuracy of this prediction rate was reported to be considerably better than what the units could achieve on their own.

Minneapolis, Minnesota: Micro Crime Hot Spots

"The little kids' playground was the gang hangout," remarked North Minneapolis Peavey Park employee Jeanie Kneath. Gang activity in the nearly 80-year-old public park had made the area a hot spot for drug use, prostitution, and other illicit activities. In the summer of 2009, a man opened fire from the playground at a group of men on the basketball courts; fortunately, the act resulted in no fatalities. As of July 2011, violent crime in Peavy Park had increased by 36 percent since 2010, despite the fact that crime rates were declining across the city.[27] For many years, the Minneapolis Police Department has studied and designed tactics to prevent this type of concentrated criminal activity.

In the 1980s, the department enlisted the help of criminologists Lawrence Sherman and David Weisburd in improving its crime analysis capabilities. In the 1990s, it officially established its crime analysis program, which has since grown into a large-scale report management system and predictive policing effort that is integrated with the department's weekly strategic planning. Department leadership reviews data-driven analysis produced by the unit in a series of what it calls CODEFOR (computer-optimized deployment–focus on results, similar to CompStat) meetings. These meetings then guide

---

[27] Matt McKinney, "Taking Back Peavey Park," *Minneapolis Star Tribune*, July 27, 2011.

the department's tactics to address predicted crime areas and times identified through statistical forecasts, pattern mapping, and an analysis of trends and data anomalies.[28]

Timely and accurate information about where crime is occurring is critical to the weekly CODEFOR meetings. During these meetings, police gather to resolve hot spots that have been identified by analysts from the most current stream of crime data. The initiative involves employees across the department—patrol officers, investigative personnel, key members of the administration, and specialized units, such as crime analysis and support services. During the CODEFOR sessions, participants focus on areas with clusters of crime, developing tactics that vary with the area in question and type of crime most prevalent. These tactics have included evidence-based approaches, such as issuing citations for jaywalking or loitering, and cracking down on drug dealing and other serious crimes.[29]

In support of CODEFOR meetings and other crime prevention efforts, the Minneapolis Police Department's Crime Analysis Unit has developed strategies for reacting to crime incidents that occur both in patterns and in micro-locations. In 2009, following a series of armed robberies of fast-food restaurants, the Crime Analysis Unit found a projected regression point within half a mile of the next actual incident, effectively forecasting the next robbery using data on the prior robberies. It has also developed evidence-based strategies to predict the criminal activities and robbery targets of specific suspects recently released from prison. This individual-focused analysis revealed that, on being released from prison, one suspect would likely rob small boutique businesses or laundries where lone female employees worked without video surveillance. The Minneapolis Police Department used the information to create patrol zones to reduce criminal opportunities in areas with high concentrations of potential victims.

According to then-Chief Tim Dolan, the data-driven strategies "paid off in North and South Minneapolis, areas that led the city last year in reducing overall crime rates." The department estimates that half of the city's most serious crimes are now concentrated in 6 percent of the regional area. As a result of this finding, the department has installed video surveillance in many of the city's highest-crime locations and directed police patrols to one- to three-block micro–hot spots. The revised tactics have helped the department respond more quickly to incidents.[30]

In 2011, Minneapolis added to its predictive policing capabilities by investing in a state-of-the-art police intelligence center, where a few officers will be posted to monitor hundreds of live video feeds from across the city. Speaking about the new resource, then-Chief Dolan said that the center would help the department send police where

---

[28] Jeff Egge, "Experimenting with Future-Oriented Analysis at Crime Hot Spots in Minneapolis," *Geography and Public Safety*, Vol. 2, No. 4, March 2011.

[29] For more information about the Minneapolis Police Department's CODEFOR, see City of Minneapolis, "What Is CODEFOR?" web page, last updated September 27, 2011.

[30] Matt McKinney, "Targeting the Next Crime," *Minneapolis Star Tribune*, January 26, 2011.

they need to be by integrating live data feeds and updated estimates of risk across the city. As of 2012, the Minneapolis Police Department's Fusion Center was housed in the First Precinct and operated continuously, responding to requests from pedestrians and law enforcement officers on the streets. This type of resource has reportedly further enhanced the city's ability to recognize and respond to critical events.

Charlotte-Mecklenburg County, North Carolina: Foreclosures and Crime

On crime, the urban environment, and the home foreclosure crisis, Wilson and Paulsen wrote that "opportunities for crime emerge, disappear, or move as the urban landscape changes."[31] Depressed neighborhoods perpetuate social decay and disorganization, which make them more amenable to criminal activity. The foreclosure crisis not only altered the social fabric between neighbors, but it also opened new avenues for criminal opportunities. Clusters of foreclosed homes left abandoned are targets for vandalism and squatting; as homes fall into disrepair, quality of life in the neighborhood declines. With fewer neighbors watching the streets and more opportunities for thieves to hide in empty homes, these communities see higher robbery and burglary rates.

In 2005, the Charlotte-Mecklenburg Police Department observed growing disorder in several neighborhoods with high numbers of foreclosed homes. The areas experienced increased juvenile delinquency, vandalism, and curfew violations. Entire communities had begun to show signs of blight and decline. The department's predictive policing program used data analysis to identify neighborhoods on the cusp of disorder, delinquency, and other causes of decline in an effort to form a community response.

After evaluating the causes and effects of the problem, the department worked with community organizations to address the increased risk of crime. Police department staff first analyzed the trends that they had observed from regular patrols by combining data from a biannual quality-of-life survey of Charlotte-Mecklenburg neighborhoods with a homeownership and sales study provided by the local newspaper to evaluate the scale of the foreclosure crisis. Visualizing the data between 2003 and 2007 with mapping tools, the department developed a series of strategies to support targeted local communities. The visuals revealed that most foreclosures had occurred in a neighborhood called Brookshire Corridor, and mostly within "affordable housing" communities that had been constructed in the previous five to seven years.[32]

The department used the information about the Brookshire Corridor communities to test other areas for their risk of foreclosure. In a test of 25 affordable housing communities, 13 high-risk neighborhoods, and 12 low-risk neighborhoods, analysts found that 96 percent of the combined-area foreclosures were in the high-risk commu-

---

[31] Ronald E. Wilson and Derek J. Paulsen, "Foreclosures and Crime: A Geographical Perspective," *Geography and Public Safety*, Vol. 1, No. 3, October 2008.

[32] Michael Bess, "Assessing the Impact of Home Foreclosures in Charlotte Neighborhoods," *Geography and Public Safety*, Vol. 1, No. 3, October 2008.

nities. Further data analysis showed that the high-risk communities consisted mostly of rental properties, and these areas also experienced higher levels of violent crime incidents than other areas.

To stabilize the neighborhoods, the department partnered with neighborhood preservation organizations. For example, one community partnership arranged for a local contractor to replace the landscaping in a troubled community. The Charlotte community also created a website with resources for preventing foreclosures. Ultimately, the Charlotte-Mecklenburg Police Department had identified a problem using geospatial data analysis and exposed the underlying community trends; it then successfully partnered with community organizations to address both components.

Crime Maps: Community Relations

The use of maps to inform the public of high-crime areas is one way predictive policing is "actioned." By creating city maps that identify high-crime areas, law enforcement is essentially reporting on the results of its predictive methods while cautioning the public to avoid these areas, another strategy to prevent crime or to prevent individuals from becoming crime victims. Although many police departments develop these visuals internally, others subscribe to commercial services, such as CrimeMapping.com, CrimeReports, and RAIDS Online, that provide a range of options and views. In this section, we briefly discuss how several cities have used crime maps to inform the public. While not explicitly "actioning" a predicted problem area, the example programs published aggregate data to citizens and key stakeholders, providing guidance on how they might modify their own behavior based on the predictions and avoid crime hot spots.

Portland, Oregon

The Portland Police Bureau's CrimeMapper, launched in 2006, is accessed by more than 1.8 million visitors each year. The web-based application makes available to the general public a suite of mapping tools that generate point-locations of crimes by type, highlight crime patterns across the city, and offer a host of other features.[33] Many cities have turned to GIS as a means of communicating to the general public the state of criminality in their communities. Clearly drawn maps are valuable for CompStat-type discussions and problem-oriented policing, and they are invaluable for neighborhood outreach.

Albuquerque, New Mexico

Local residents are generally cognizant of which alleys, street corners, and neighborhoods to avoid, but crime maps can help reveal the other risks and larger, less visible threats to a community. Drinking and driving is a national problem, but without a heavy police presence, regular sobriety stops, and frequent public announcements, the

---

[33] Nora Parker, "Portland Police Bureau Makes Geospatial Widely Accessible," *Directions Magazine*, February 11, 2008.

telltale signs of this epidemic may be hard to identify in some communities. A 2005 survey of citizens in Albuquerque, New Mexico, revealed an awareness that drunk driving was a significant problem in the city, but few believed that the problem was pulsing through their neighborhoods.[34] Although traffic accidents were not uncommon in Albuquerque, the general public had no way of distinguishing between an ordinary traffic accident and one involving alcohol.

The DWI Resource Center was founded in Albuquerque to reduce the social and economic costs associated with driving while intoxicated (DWI) through research and outreach. In 2005, the DWI Resource Center developed a sequence of traffic accident maps that overlaid traffic congestion data with accidents involving alcohol in an effort to counter inaccurate public perceptions of where these crimes occurred. One particularly telling map demonstrated that 64 percent of accidents between 7:00 p.m. and 5:59 a.m. involved alcohol. This research helped police and community leaders identify target areas for more DWI checks.

Cincinnati, Ohio

As in Albuquerque, maps can be used not only to correct public perceptions but also to develop location-specific law enforcement and prevention strategies. Blight is a common indicator of economic decline, and strategies to address this problem are far from simple. In 2008, the Cincinnati Neighborhood Enhancement Program (CNEP) was recognized as the best physical revitalization program of the year by the national nonprofit organization Neighborhoods USA. The goal of the program is to develop neighborhood assets through a series of concentrated 90-day revitalization campaigns. Neighborhoods are identified through a geospatial analysis of crime patterns, disorder-related calls for service, a prevalence of vacant buildings, abandoned cars, and the presence of litter and weeds.[35]

Once the target areas of the city are identified, the CNEP team of city personnel, community leaders, and volunteers assemble a list of specific revitalization projects based on the location-specific data and environmental analysis. Example improvements include landscaping and other beautification efforts, improved lighting, and barriers to defend the public right of way. Other strategies to tackle blight might be concentrated on building-code enforcement, targeted traffic enforcement, drug and disorder policing, or other approaches to "cool down" crime hot spots. Because CNEP's goal is to make a difference in a troubled community in just 90 days, the strategies must be designed in a way that allows for quick execution. Since CNEP's implementation, the city of Cincinnati has reported an overall decrease in crime and a 20-percent

---

[34] Tom Beretich, "Mapping Programs Target Alcohol-Impaired Driving," *Geography and Public Safety*, Vol. 1, No. 2, July 2008.

[35] City of Cincinnati, "Neighborhood Enhancement Program," web page, undated.

increase in property value.[36] While these data have not been evaluated scientifically, this case study provides a good example of how data and insights from law enforcement have informed broader community revitalization initiatives that can ultimately reduce crime rates.

Hartford, Connecticut

Although maps can be powerful tools for communicating neighborhood information to citizens and government agencies, some organizations have a greater capability to use the resource than do others. In 1997, Hartford, Connecticut, was one of 12 cities to be awarded a Comprehensive Communities Program grant by the U.S. Department of Justice, Bureau of Justice Assistance. At the time, Hartford already had a demonstrated history of neighborhood-supported and citizen-based problem-solving policing efforts. With the grant funding, the city formed 17 formal community groups and provided them with location-specific crime information through a city-created program called Neighborhood Problem Solving (NPS). The computer-based NPS system relayed information on crime times, locations, and times but omitted names and other personal details.[37]

Thomas Rich of Abt Associates studied the Hartford implementation of the NPS system and the execution of the grant on behalf of NIJ by assessing how each community group used the NPS data. One major finding was that the community organizations varied in size and level of support. One organization had as many as 15 people, with a few full- and part-time paid staff and a group of active volunteers. Other organizations had one or two paid personnel, and still others were merely groups of concerned citizens who met occasionally to review the data. The information was also used for a variety of different purposes, ranging from improving street lighting to publicizing job fairs or delivering city services to specific residents. Although a few of the 17 groups did not regularly use the NPS data, the most common use was to generate a map that would spur discussion and increase awareness of community problems.

Helping to raise neighborhood awareness about an issue is an important use of GIS data. Such uses raise privacy concerns, however. The U.S. Department of Justice has released a report exploring some of these issues and offering guidance on sharing and using crime maps and spatial data.[38]

---

[36] Anthony A. Braga and David L. Weisburd, *Policing Problem Places: Crime Hot Spots and Effective Prevention*, New York: Oxford University Press, 2010.

[37] Thomas Rich, "Crime Mapping and Analysis by Community Organizations in Hartford, Connecticut," *National Institute of Justice Research in Brief*, March 2001.

[38] Maps are "conversation starters" and useful for a variety of neighborhood-oriented policing strategies, but privacy and public use are concerns that should not be overlooked. Wartell and McEwen outline several considerations when designing crime maps, including personal privacy, the social impact of providing crime maps, and the need to protect the data from benefiting criminals. See Julie Wartell and J. Thomas McEwen, *Privacy in the Information Age: A Guide for Sharing Crime Maps and Spatial Data*, Washington, D.C.: National Institute of Justice, July 2001.

Police Actions

This chapter focused on the police intervention portion of the prediction-led policing business process and, in some cases, criminal responses. The Center for Evidence-Based Crime Policy at George Mason University has concluded that police can be most effective in reducing violent crime when they are proactive. The examples and case studies illustrate this to some extent. The Columbia Heights example, while hypothetical, showed how intervention at the problem-specific level can be more effective than declaring an entire neighborhood a hot spot (generic area intervention).

The Sacramento Police Department's test of Koper's 13- to 15-minute stay rule resulted in a rather dramatic decrease in Part 1 crimes while avoiding crime displacement to other areas. In this case, problem-specific intervention led to a favorable criminal response. In Memphis, Blue CRUSH points police toward specific high-risk locations (street corners and city blocks) and times. These locations get the most attention. In this case, problem-specific intervention was effective in reducing crime at specific locations.

In all the cases presented, the police were proactive and intervened in some way based on information obtained from some combination of predictive methods and traditional police intelligence. In the next chapter, we discuss the problems associated with identifying "hot people" (i.e., perpetrators and suspects).

CHAPTER FOUR

Using Predictions to Support Investigations of Potential Offenders

Round up the usual suspects.

—Captain Louis Renault at the end of Casablanca

Chapter Two addressed predictive techniques, principles, and the capacity in place at some police departments to estimate *where* and *when* a crime might occur and its likely causes. Chapter Three reviewed police interventions based on these estimates and other sources. The focus was clearly on the crime. This chapter furthers that discussion to address how predictive technology has been used to estimate who—who is most likely to commit crimes in the future or who most likely committed crimes in the recent past. The methods discussed in Chapter Two built on crime incident records with information on temporal and spatial parameters. This chapter adds the behavioral dimension to the equation. For example, data mining of behavioral patterns has been used to match young offenders in the Florida juvenile justice system to programs that are designed to reduce recidivism rates based on the outcomes of statistically "similar" youth. In this case, the ability to predict has been applied to prevent criminal behavioral patterns by assigning a young offender to the right intervention. In other cases, departments have used computer-assisted searches of criminal intelligence and external databases to identify the most likely suspects in crimes of interest.

There are two important caveats involving predictions regarding potential offenders. The first is that, compared with predictions related to spatiotemporal crime techniques, methods for making predictions involving people are much less mature. The second is that privacy and civil rights considerations are paramount, so we begin with a discussion of these issues.

Protecting Privacy Rights and Civil Liberties

Assessing the "risk" associated with an individual—whether of committing future crimes or of being a suspect in past crimes—is highly contentious and fraught with personal privacy concerns. The phrase *predictive policing* began appearing in headlines shortly after the release of the Tom Cruise movie *Minority Report*, which insinuated that the police and other law enforcement authorities have the knowledge and power to arrest (or at least detain) suspects before they have committed a crime.[1] Headlines referring to predictive policing technology as a "real-life *Minority Report*" gave the impression that portions of the movie were based on actual law enforcement practices in development. In the movie the public was seemingly deprived of any right to privacy at all, a price paid for a crime-free society. As "predictive policing" was thrust into the headlines, many citizens, including privacy rights activists, likened the phrase to a significant loss of privacy in their lives.

Predictive Policing Symposium Assessment

Privacy was one of three primary topics of discussion at NIJ's first predictive policing symposium, held in Los Angeles in November 2009.[2] Participants acknowledged that privacy and civil liberties were critically interrelated with predictive policing, noting that constitutionality would be a requirement in ensuring a solid foundation for the future use of these approaches. They agreed that engaging privacy advocates and community leaders would be critical to successful implementation. Participants also cited transparency, auditing, and due diligence as key elements of predictive policing. The discussion emphasized departmental responsibility to approach concerns about privacy and civil liberties and promoted regular audits and other practices to ensure that systems are operated with due diligence.

A distinction between *intelligence* and *information* also emerged from the discussion.[3] Participants said that, at the time, they saw a lack of clarity in the concept of predictive policing and that this could lead to confusion about which types of intelligence are valued and which information activities could lead to legal challenges. The symposium group on privacy and civil liberties concluded with several key points:

1. Policing has a rich history of dealing with privacy and related mistakes, and these issues have yet to be resolved.

---

[1] The film, based on the sci-fi novella by Philip K. Dick, was released in 2002. The story is set in 2054 and revolves around the activities of a recently established law enforcement agency known as PreCrime. The agency employs three psychics called "precogs," who have the ability to predict crimes and suspects. The result is a crime-free city, but with it comes the ethically questionable practice of arresting people for crimes they have not (yet) committed or—if the precogs' predictions are fallible—arresting people for crimes they may never have committed.

[2] National Institute of Justice, "Predictive Policing Symposium: The Future of Prediction in Criminal Justice," web page, last updated December 18, 2009.

[3] The United Nations manual on criminal intelligence states, "In its simplest form, intelligence analysis is about collecting and utilizing information, evaluating it to process it into intelligence, and then analysing that intelligence to produce products to support informed decision-making" (United Nations Office on Drugs and Crime, *Criminal Intelligence: Manual for Analysts*, Vienna, Austria, April 2011). Simply put, intelligence is evaluated information.

2. The suspicious activity reporting effort provides lessons for how to develop a privacy policy.[4]
3. There will come a time when training in privacy issues is considered as important to a policing program as its firearms policy.
4. Transparency is critical to establishing community trust.
5. Understanding what behaviors have a nexus to crime provides a valid purpose for law enforcement.
6. Predictive policing must be constitutional.

Craig Uchida, an active participant at the symposium, summarized these discussions and conveyed the challenge of protecting civil liberties as a call for departments to keep their processes transparent. He described a perceived public fear about predictive policing: "The term raises fears that police might engage in illicit tactics—that they will overstep their bounds and potentially use information and intelligence in a way that abridges the Constitution."[5] This concern necessitates an examination into the types of information that are protected and the mechanisms (e.g., publishing, legal) by which that information can be shared for predictive initiatives. Uchida emphasized regular communications with the public about the insight and range of activities stemming from predictive policing. This type of public outreach is necessary for departments to ensure acceptance within their communities and in supporting partnerships with the public. Speaking about partnership and policing, former Los Angeles Chief Bill Bratton added that, "when we reduce the image of the police doing it to the community, it becomes the community doing it with the police."

Privacy Under the Fourth Amendment of the U.S. Constitution

Discussions of this nature—about the balance between personal privacy and the state's need to maintain public safety—are not new, but they have become increasingly important with advances in technology for data collection, processing, and analysis. The trails of personal information left in a person's wake are nearly endless—emissions ranging from the DNA in a skin flake to the digital exhaust willingly or unknowingly produced by cell phones, computers, and other technologies. How that personal information is collected and used by a third party is a concern for leaders and legal scholars in the policing community.

The Fourth Amendment is the primary constitutional limit on the government's ability to obtain personal or private information. It prohibits unreasonable searches and seizures according to a series of tests that have been defined by case law. The most

---

[4] For a list of privacy resources, see Institute for Intergovernmental Research, "Nationwide SAR Initiative: Resources," web page, undated.

[5] Craig D. Uchida, *A National Discussion on Predictive Policing: Defining Our Terms and Mapping Successful Implementation Strategies*, prepared for the National Institute of Justice, Document No. NCJ230404, May 2010.

recent addition to this rich case law is the 2012 Supreme Court case *United States v. Jones*, which addressed the government's installation and prolonged use of a Global Positioning System (GPS) tracking device on a suspect's car without a warrant. The court found that the police installation of a GPS device without a warrant constitutes a "search" under the Fourth Amendment, and the case established the "trespass-based" test to use when determining the legality of a potential police action.[6]

The Fourth Amendment provides little to no protection for data that are stored by third parties, however. The Supreme Court case *United States v. Miller* held that there is no reasonable expectation of privacy when it comes to information held by a third party.[7] The legal question considered by *United States v. Miller* focused on the personal information written onto canceled checks that are then handled by banks and bank employees. Reviewing the statutory considerations, Stephanie Pell considered different forms of personal data held by third-party organizations that may be used for policing purposes: communications, financial information, and other material not covered by other laws.[8] Pell concluded that the government can request, with a subpoena or national security letter, some third-party data that are not already protected by the U.S. Constitution or by another statute if these data are not already made available voluntarily by the third party. Pell describes this "lack of regulation" when it comes to third-party data as the potential basis for public-private partnerships similar to those the government maintains with major banks, airlines, and other organizations to protect against national security threats (e.g., identifying people who are financing terrorism or in the stages of planning a destructive act).[9]

The lessons learned through *United States v. Jones* clearly define a class of intelligence-related policing tactics that would not be permitted under the law. However, data used for predictive policing purposes may also come from partnerships with third-party data collectors, and, in this case, training on the implications of *United States v. Miller* is important. The legal framework for personal data protection in the era of "big data" is still being defined, and personal privacy protections are not yet clear in jurisprudence or in public perceptions.

Privacy with Respect to Policing Intelligence Information Systems

The system design parameters and considerations for sharing data across police jurisdictions while protecting privacy rights have been discussed since the 1980s, and these discussions will remain relevant for departments planning predictive policing initia-

---

[6] *United States v. Jones*, U.S. Supreme Court Docket No. 10-1259, 2012.

[7] *United States v. Miller*, 425 U.S. 435, 1976.

[8] Stephanie K. Pell, "Systematic Government Access to Private-Sector Data in the United States," *International Data Privacy Law*, Vol. 2, No. 4, 2012.

[9] Fred H. Cate and Beth E. Cate, "The Supreme Court and Information Privacy," *International Data Policy Law*, Vol. 2, No. 4, 2012.

tives. This literature and various governing policies provide a strong legal framework for developing and discussing the potential for new systems. In the 1980s, the first criminal intelligence system operating policies were established, primarily at the U.S. federal level, to expand the legal considerations for how data related to narcotics operations would be handled by law enforcement agencies.[10] The policies were initially applied to the Regional Information Sharing Systems program but were later expanded to cover the Organized Crime Narcotics project and other intelligence database operations funded by the U.S. Department of Justice. All criminal intelligence systems that operate with funding specified by the Omnibus Crime Control and Safe Streets Act of 1968 are required to conform to these policies, which outline the privacy and constitutional rights of citizens.[11]

In the wake of the September 11, 2001, terrorist attacks, the National Criminal Intelligence Sharing Plan (NCISP) was established by a collective of leaders from across the policing community. The NCISP recommended that all states voluntarily adopt the policies from Title 28, Part 23, of the Code of Federal Regulations for taking privacy into account when developing or using any intelligence system.[12] The NCISP also recommended that state agencies consider the Association of Law Enforcement Intelligence Units' (LEIU's) Intelligence File Guidelines as a model for maintaining intelligence data files. Marilyn Peterson provides an excellent summary of the policies in these two documents. The following bullets are taken from the 2005 U.S. Department of Justice report titled *Intelligence-Led Policing*:

- Information entering the intelligence system should meet a criminal predicate or reasonable suspicion and should be evaluated to check the reliability of the source and the validity of the data.
- Information entering the intelligence system should not violate the privacy or civil liberties of its subjects.
- Information maintained in the intelligence system should be updated or purged every five years.
- Agencies should keep a dissemination trail of who received the information.
- Information from the intelligence system should be disseminated only to those personnel who have a right and need to know in order to perform a law enforcement function.[13]

---

[10] Code of Federal Regulations, Title 28, Judicial Administration, Part 23.30, Operating Principles.

[11] Public Law 90-351, Omnibus Crime Control and Safe Streets Act, June 19, 1968; see Marilyn Peterson, *Intelligence-Led Policing: The New Intelligence Architecture*, Washington, D.C.: U.S. Department of Justice, Bureau of Justice Assistance, Report No. 210681, September 2005.

[12] U.S. Department of Justice, *The National Criminal Intelligence Sharing Plan*, Washington, D.C., October 2003.

[13] Peterson, 2005, p. 20.

Intelligence information sharing across law enforcement agencies has become a critical component of modern public safety operations. These operations rely on data, and law enforcement agencies must develop and maintain clear policies for handling those data that are consistent with federal, state, and other regulations. Title 28, Part 23, of the Code of Federal Regulations and the LEIU Intelligence File Guidelines are two of the best-known policies for managing privacy concerns through intelligence data systems. *The Police Chief* magazine published a special feature on information sharing in 2006 with a review of the implementation and privacy considerations for the National Data Exchange (N-DEx), developed in 2005.[14] The N-DEx was designed to enable data sharing between agencies across jurisdictional boundaries and to provide a new series of analytic tools to aid in investigations of criminal and terrorist threats. The article highlights several aspects of the system's development as critical to its early success; a few of those related to privacy concerns include the following:

- *Clear ownership of data elements:* The law enforcement agency that submits data retains ownership and control over its data. Thus, the program clearly defines the system controls and centralizes responsibility for policies governing data dissemination and privacy.
- *Clear standards for classification:* Only data classified as sensitive but unclassified or below are permitted in the N-DEx system. Protocols for classification helped avoid legal conflicts related to restricted-data spills. Developers integrated additional support for case sensitivity information o allow departments to restrict certain cases and data from their records management system (RMS) for various privacy and case sensitivity reasons.

These and other privacy solutions have been cited as important factors in the successful implementation of the N-DEx, as formalized protocols addressing privacy and case sensitivity were baked into the information system architecture. Intelligence gathering and the pooling of intelligence resources across regional jurisdictions have helped made law enforcement agencies more effective. However, these activities have also led to considerable privacy concerns. As the technology for information sharing advances, clear and consistent use of the policies outlining protections for privacy will remain a law enforcement responsibility.

Privacy Resources for the Law Enforcement Community
A range of tools are in development that assess a person's risk of committing or observing a crime based on certain behavioral patterns, but police departments will want to

---

[14] Mark A. Marshall, "N-DEx: The National Information Sharing Imperative," *The Police Chief*, Vol. 73, No. 6, 2006.

proceed with caution. Law enforcement officials must carefully consider issues related to privacy and First Amendment rights before acting on information of this nature.

Proper management of sensitive data is a significant concern, and improper use can have legal ramifications as well as personally damaging consequences for citizens and communities. A police department's best protection against the improper collection or use of private personal data is to understand the tests associated with the Fourth Amendment of the U.S. Constitution and other statutes related to privacy. Technology advances are regularly testing the bounds of these precedents; police departments will need to maintain an open channel of communication with their legal teams for help in navigating challenges to a citizen's right to privacy.[15] Title 28, Part 23, of the Code of Federal Regulations and LEIU Intelligence File Guidelines will be useful resources as police departments form information sharing partnerships across jurisdictions with other law enforcement agencies.

In this section of the guide, we do not cover the many state and local government policies that protect a citizen's right to privacy. Many privacy laws in the United States, Canada, Europe, and other regions are structured around the widely accepted Fair Information Practice Principles. These principles were first established by the U.S. Department of Health, Education, and Welfare in 1973. The following excerpts are from the Organisation for Economic Co-operation and Development's Guidelines on the Protection of Privacy and Transborder Flows of Personal Data, which were adopted on September 23, 1980:

Openness Principle
There should be a general policy of openness about developments, practices and policies with respect to personal data. Means should be readily available for establishing the existence and nature of personal data, and the main purposes of their use, as well as the identity and usual residence of the data controller.

Collection Limitation Principle
There should be limits to the collection of personal data and any such data should be obtained by lawful and fair means and, where appropriate, with the knowledge or consent of the data subject.

Purpose Specification Principle
The purposes for which personal data are collected should be specified not later than at the time of data collection and the subsequent use limited to the fulfillment of those purposes or such others as are not incompatible with those purposes and as are specified on each occasion of change of purpose.

---

[15] Jennifer Valentino-DeVries, "How Technology Is Testing the Fourth Amendment," *Wall Street Journal*, September 21, 2011.

Use Limitation Principle
Personal data should not be disclosed, made available or otherwise used for purposes other than those specified . . . except a) with the consent of the data subject; or b) by the authority of law.

Data Quality Principle
Personal data should be relevant to the purposes for which they are to be used, and, to the extent necessary for those purposes, should be accurate, complete, and relevant and kept up-to-date.

Individual Participation Principle
An individual should have the right: a) to obtain from a data controller, or otherwise, confirmation of whether or not the data controller has data relating to him; b) to have communicated to him, data relating to him within a reasonable time; at a charge, if any, that is not excessive; in a reasonable manner; and in a form that is readily intelligible to him; c) to be given reasons if a request is denied and to be able to challenge such denial; and d) to challenge data relating to him and, if the challenge is successful, to have the data erased, rectified, completed or amended.

Security Safeguards Principle
Personal data should be protected by reasonable security safeguards against such risks as loss or unauthorized access, destruction, use, modification or disclosure of data.

Accountability Principle
A data controller should be accountable for complying with measures which give effect to the principles stated above.[16]

The State of California has developed privacy laws based, in part, on the Fair Information Practice Principles, and it has used the framework to develop many state-specific protections for its citizens.[17] The purpose of this introduction to some of the privacy considerations related to predictive policing initiatives is as much to inform as it is to caution that the applicable regulations and guidelines are numerous. However, this also means that there are many resources and communities of interest for law enforcement agencies to consult. To support the policing community, NIJ and others have published a range of resources on data sharing and privacy concerns. The following are a few better-known examples:

---

[16] Organisation for Economic Co-operation and Development, "OECD Guidelines on the Protection of Privacy and Transborder Flows of Personal Data: Background," September 23, 1980.

[17] The California Department of Justice hosts an exhaustive list of the privacy rights California citizens enjoy. See California Department of Justice, Privacy Enforcement and Protection Unit, "Privacy Laws," web page, undated.

- National Criminal Justice Association, *Justice Information Privacy Guideline: Developing, Drafting and Assessing Privacy Policy for Justice Information Systems*, Washington, D.C., September 2002; supported by a grant from the Bureau of Justice Assistance
- Marilyn Peterson, *Intelligence-Led Policing: The New Intelligence Architecture*, Washington, D.C.: U.S. Department of Justice, Bureau of Justice Assistance, Report No. 210681, 2003
- Julie Wartell, and J. Thomas McEwen, *Privacy in the Information Age: A Guide for Sharing Crime Maps and Spatial Data*, Washington, D.C.: National Institute of Justice, July 2001.

Dealing with Noisy and Conflicting Data: Data Fusion

We introduced the concept of data fusion in Chapter One. Data fusion is a process used to combine elements of incoming data to improve the accuracy of an assessment. When dealing with information about people—which are especially likely to be noisy and conflicting—law enforcement agencies might benefit from a formal process to address data noise and confusion. We have found little evidence of such processes in the law enforcement community, however. The combining methods briefly described here offer a starting point as agencies attempt to make sense of data on individuals and groups.

Heuristic and Simple-Model Methods

Heuristic and other simple data fusion methods include checklists and risk indexes, which are especially suitable for on-the-scene law enforcement personnel at check-points. They are, perhaps, less so for a police department's analysis section. Checklists are already common and can be either negative (i.e., any indicator, if met, triggers additional screening) or positive (i.e., if all indicators are met, secondary screening can be minimized). Index or scoring methods typically characterize a risk level by summing indicator scores or by computing *risk* as the product of a *likelihood* and a *consequence*, with a score exceeding a threshold triggering additional screening. Significantly, good scoring methods sometimes need to be nonlinear and should be empirically validated rather than ad hoc. In this chapter, we also consider more complex "simple" methods, such as scorecards and conditional-indicator sets. Most of the instruments used to assess personal and group risk that we discuss here fall into this category.

More Sophisticated Fusion Methods

More advanced fusion methods are likely necessary in using behavioral indicators (behavior usually associated with criminal behavior) for law enforcement purposes, particularly at the level of a department's analysis unit. Such methods are better equipped to reduce the often-serious signal-to-noise and false-alarm problems that accompany information about people. There are several suitable mathematical techniques for this purpose, but each comes with some drawbacks. Bayesian updating is now well understood and widely applied in other domains, but its usefulness in law enforcement is limited by its demands for many subjective estimates of conditional probabilities, for which there is and will continue to be an inadequate base.[18] Some relatively new methods are based on Dempster-Shafer belief functions, which distinguish between having evidence *for* a proposition (such as an individual's malicious intent) and having *contrary* evidence (of innocence, in this example). Both can be high, whereas if the language used were that of simple probabilities, a high probability of malicious intent would imply a low probability of innocence.[19] Ultimately, there are several major shortcomings in using this approach as well. A much newer approach, called the Dezert-Smarandache Theory, has not yet been widely examined or applied, but something along these lines has promise for law enforcement. The theory deals specifically with combining evidence from sources that produce imprecise, fuzzy, paradoxical, and highly conflicting reports—precisely the type of reports encountered in law enforcement. For example, it would allow an analyst to characterize evidence that (1) both John and Harry belong to a criminal gang, (2) either John or Harry (but not both) belongs to the gang, and (3) John belongs to the gang or John does not belong to the gang.[20] Other available methods include "possibility theory," various multiattribute theories, "mutual information" (which builds on the concept of information entropy), and Kalman filtering.

The best data management methods for law enforcement are not yet certain, but it appears that some of these methods might help in building a case for monitoring a given individual or group more closely. Some of the more sophisticated mathematical methods are not likely to be universally appropriate, but simple techniques, such as checklists and indexing are within the purview of most departments. This is an area where further research might be beneficial.

Risk Assessment for Individual Criminal Behavior

Can we estimate the risk that a person will commit a serious crime in the near future? How well will a person respond to a short-term change in life circumstances (e.g.,

---

[18] See, for example, Howard Raïffa, *Decision Analysis: Introductory Lectures on Choices Under Uncertainty*, Boston, Mass.: Addison-Wesley, 1968.

[19] For the definitive work on the Shafer-Dempster method, see Glenn Shafer, *A Mathematical Theory of Evidence*, Princeton, N.J.: Princeton University Press, 1976.

[20] See Florentin Smarandache and Jean Dezert, eds., *Advances and Applications of DSmT for Information Fusion (Collected Works)*, Rehoboth, N.M.: American Research Press, 2009.

getting fired from a job, fighting with a spouse or partner, abusing alcohol or drugs)? What is the likelihood that these short-term circumstances will motivate a person toward criminal behavior? A 1996 study led by Julie Horney and sponsored by NIJ attempted to capture the social life events that may tip the scale for a potential offender in the direction of criminal activity. The research team collected data on the month-to-month changes in the circumstances that shaped the life of people recently arrested, capturing the two-year period that preceded the arrest. The survey was administered to 658 newly convicted male offenders serving sentences in the Nebraska Department of Correctional Services. The findings suggest that short-term negative changes in life circumstances may sharply increase criminal activity—for example, the use of illegal drugs increased the odds of committing property crime by 54 percent and committing an assault by more than 100 percent.[21] In contrast, such positive short-term changes as living with a girlfriend, attending school, or receiving justice system supervision may decrease the odds of recidivism.

This section introduces the behavioral instruments that are most commonly used by the correctional system and describes the challenges and considerations involved in employing them.

Commonly Used Behavioral Instruments

A range of techniques have been developed to assess behavioral patterns. Tools to assess risk among potential offenders include the Level of Service Inventory–Revised (LSI-R) assessment, a quantitative survey tool used by parole officers to determine the level of supervision and treatment typically required by a person going on parole. According to Dawn Clausius, a police intelligence analyst with the Olathe, Kansas, Police Department, these assessments may hold "mountains of untapped data for predictive policing efforts."[22] Behavioral patterns from the LSI-R might be integrated with geospatial and temporal data on criminal dynamics in a region to more robustly focus on areas frequented by recently paroled offenders with a higher than average risk level. In fact, some police jurisdictions are using this information to identify the parolees who will receive both parole and police attention during their supervision. Combining the two resources to closely monitor these high-risk offenders would help police predict future behavior while parole officers monitor for possible violations that might result in parolees' return to custody.

The LSI-R is among the instruments most frequently used by corrections agencies for classifying and assessing risk levels of offenders. Essentially, it attempts to capture

---

[21] Julie Horney, D. Wayne Osgood, and Ineke Haen Marshall, "Criminal Careers in the Short-Term: Intra-Individual Variability in Crime and Its Relation to Local Life Circumstances," *American Sociological Review*, Vol. 60, No. 5, October 1995.

[22] Brian Heaton, "Behavioral Data and the Future of Predictive Policing," *Government Technology*, November 2, 2012.

an offender's background. The instrument includes 54 items that are divided into ten subscales; the total number of checked items is equal to the total score on the LSI-R. The higher the score, the greater the risk of criminal behavior. The subscales cover the following ten areas: criminal history, education and employment, finances, family and marital conditions, accommodation, leisure and recreation activities, companions, alcohol and drug problems, emotional and personal issues, and attitudes and orientations. Generally, a score of 29 or higher represents the maximum risk level, a score between 19 and 29 is thought to equate to medium-level risk, and a score of 18 or below is considered an indication of minimal risk.

The ten sub-scales also provide a means to classify the degree of *static* versus *dynamic* risk factors in relation to criminal behavior and recidivism. Static risk factors are conditions in an offender's past and are not responsive to correctional intervention, whereas dynamic factors are conditions that can be changed by programs, treatment, counseling, and other interventions. A static risk factor may include mental or physical limitations; dynamic factors could be conditions related to employment, education, or associations. These factors may be used to classify offenders for specific intervention programs that aim to minimize the chance of recidivism.[23]

Other tools used by the corrections system to assess and classify offenders include the Substance Abuse Subtle Screening Inventory (SASSI), the Beck Depression Inventory–II (BDI-II), the Minnesota Multiphasic Personality Inventory–2 (MMPI-2), the Addictions Severity Index (ASI), the Hare Psychopathy Checklist–Revised (PCL-R), and the Michigan Alcoholism Screening Test (MAST).[24] These instruments help screen for a range of risks and behaviors associated with recidivism and criminal behavior.

A more recent development is to use predictive analytics to build formal statistical models assessing the probability that an individual will offend based on the presence of specific risk factors. For example, criminologist Richard Berk has recently developed a model that assesses the risk that an offender will commit homicides while out on parole or probation. In the model, the age at which the first major offense was committed was found to be the biggest factor, with offenders committing serious crimes at earlier ages being at greatest risk. The model can reportedly identify only about eight future killers out of 100, however.[25]

---

[23] Key Sun, *Correctional Counseling: A Cognitive Growth Perspective*, 2nd ed., Burlington, Mass.: Jones and Bartlett Learning, 2003, pp. 25–44.

[24] The SASSI and ASI can be found in John P. Allen and Veronica B. Wilson, eds., *Assessing Alcohol Problems: A Guide for Clinicians and Researchers*, 2nd ed., Bethesda, Md.: National Institute on Alcohol and Alcoholism, 2003, pp. 591–593 and 245–247, respectively. The BDI-II, MMPI-2, and PCL-R are available for purchase through Pearson Education. The MAST inventory is available in Edson Hirata, Osvaldo Almeida, Rossana R. Funari, and Eva L. Klein, "Validity of the Michigan Alcoholism Screening Test (MAST) for the Detection of Alcohol-Related Problems Among Male Geriatric Outpatients," *American Journal of Geriatric Psychiatry*, Vol. 9, No. 1., Winter 2001.

[25] Kim Zetter, "U.S. Cities Relying on Precog Software to Predict Murder," *Wired*, January 10, 2013.

Limitations of Behavioral Instruments

Although the LSI-R and a few of the other behavioral instruments mentioned here have been used to predict recidivism, several research studies have brought to light a number of considerations and challenges to better define the proper and improper use of the instruments. The primary two challenges to the LSI-R are also relevant for reviewing the effectiveness of other similar behavioral instruments: inter-rater reliability and mis-specification of the predictive model, which would influence the statistical association between the behavioral instrument and criminal outcomes, such as recidivism.

Inter-rater reliability is the degree of agreement among raters. In the context of the LSI-R and other behavioral instruments, high inter-rater reliability indicates that different trials of the tool to capture risk-level characteristics of the same population of offenders would return the same results. Low inter-rater reliability indicates that various raters do not agree and that the scale is defective or the raters need to be retrained. Implementation of the LSI-R for assessment and classification purposes has been evaluated for correctional systems, however.

An evaluation for the Pennsylvania Board of Probation and Parole reported in 2003 that most of the LSI-R scoring items did not meet a sufficient level of reliability as implemented by corrections staff. Specifically, 18 of the 54 total items (33 percent) had reliability scores at or above a minimum 80-percent threshold for inter-rater consistency; the items with the highest levels of agreement were the fact-based measures of the prisoner's criminal history. A second test for reliability on the LSI-R, completed roughly two years later, showed that reliability had improved with regard to calculating offenders' overall risk level (e.g., high, medium, low)—from 71 percent agreement among raters in 2000 to 88 percent agreement in 2002. At the item level, 63 percent of the items met a minimum threshold for inter-rater consistency. Despite these noted improvements, the evaluation report suggested that the LSI-R was not a sufficient method for assessing risk at parole interviews. The research team recommended that the Pennsylvania Board of Probation and Parole consider another behavioral instrument related to the LSI-R, called the Level of Service Inventory–Screening Version (LSI-SV).[26]

Considering the LSI-R and two other similar instruments (the Historical Clinical Risk Management–20 [HCR-20] and the PCL-R) in the context of a German prison sample, Klaus-Peter Dahle reported much higher rater reliability than the Pennsylvania but identified other important limitations to the behavioral instruments that limit the predictive accuracy of risk assessments. The primary limitations include high percentages of criminals with medium scores, which correspond to an ambiguous assessment. Additionally, predictions of reoffense achieved much more significant results when

---

[26] James Austin, Dana Coleman, Johnette Peyton, and Kelly Dedel Johnson, *Reliability and Validity Study of the LSI-R Risk Assessment Instrument*, Washington, D.C.: Institute on Crimes, Justice and Corrections, George Washington University, January 9, 2003.

simple factors about the criminal's background (e.g., demographic, criminological, and psychopathological characteristics of the offender) were incorporated into the models in addition to behavioral factors. When explanatory variables are omitted, the model suffers from misspecification, confounding the statistical significance of the predictions. Due to these limitations, Dahle suggests that the behavioral instruments should complement a series of other carefully and clinically informed appraisals and should not be used as a substitute for them when making an assessment about a prisoner.[27]

James Austin has published a useful paper on proper and improper uses of risk assessments for the corrections community that is applicable to the use of behavioral instruments for law enforcement purposes. According to Austin, the accuracy of a behavioral instrument is dependent on its *reliability* and *validity*. These dual requirements for accuracy can be evaluated with commonly used statistical tools that assess inter- and intra-rater reliability on surveys and psychometric instruments. To assess an instrument's validity, it is also useful to employ statistical methods for examining associations between variables captured by an instrument and an outcome behavior (e.g., correlation or regression-based techniques). Although behavioral instruments like the LSI-R may have been developed over many decades, failure to properly train raters and continuously consider the statistical accuracy of assessments can lead to improper assessments of risk.[28]

Austin also raises concern about the *perceived rigidity* of these "actuarial" behavioral assessments. Risk assessments are not "foolproof," and, like Dahle, Austin proposes that supplemental information that is not factored into the scoring system be incorporated to allow for adjustments on a case-by-case basis. He emphasizes the importance of caution, trained raters, and an understanding of the statistical tools to evaluate risk assessments when using behavioral instrument data for classification and predictive purposes.

Finally, Austin notes that behavioral instruments are designed with specific populations and their demographic characteristics in mind. Determining the *applicability* of a specific behavioral assessment to a location, community, and crimes of interest is an important preliminary step to integrating these tools into an investigation or predictive policing effort. Assessing the applicability of a given instrument should start with a strong understanding of the control and treatment populations used to determine the reliability and validity of the measure. With an understanding of the clinical study parameters that were used to create the behavioral instrument, law enforcement agen-

---

[27] Klaus-Peter Dahle, "Strengths and Limitations of Actuarial Prediction of Criminal Reoffence in a German Prison Sample: A Comparative Study of LSI-R, HCR-20 and PCL-R," *International Journal of Law and Psychiatry*, Vol. 29, No. 5, September–October 2006.

[28] James Austin, "The Proper and Improper Use of Risk Assessment in Corrections," *Federal Sentencing Reporter*, Vol. 16, No. 3, February 2004.

cies will be better prepared to select the tools that are most appropriate and applicable to the criminal element in their community.

Quebec, Canada: Assessing Criminogenic Risks of Gang Members

A great threat to public safety in Quebec is the presence of street gangs. Juvenile prostitution networks, drug trafficking, and other criminal activities pursued by street gangs increased the flow of juvenile offenders into the adult correctional system in Quebec. Research conducted by Jean-Pierre Guay and reported in 2012 examined the applicability of the Level of Service/Case Management Inventory (LS/CMI) to identify specific criminogenic needs profiles of gang members as compared to non–gang members.[29]

Similar to the LSI-R, the LS/CMI is an assessment that measures the risk and need factors of late adolescent and adult offenders. A refinement of the LSI-R (reducing the former 54 question survey to 43 items), the LS/CMI tool consists of 11 sections and can be used to estimate a level of risk (both static and dynamic) associated with an offender's profile.[30] Guay selected a sample of 172 offenders serving sentences of more than six months under provisional jurisdiction (86 recognized gang members and 86 non–gang members). All participants were assessed with the LS/CMI.

The LS/CMI results showed that gang members present more significant criminogenic risks and needs, and in a greater number of the sub-scale areas than did the control group non–gang members. The noticeable differences suggest that gang members experience a significantly higher level of risk and therefore have greater needs in terms of intervention. Guay discusses cognitive behavioral programs that follow Risk Need, and Responsivity (RNR) principles, proposing that successfully implemented these could reduce recidivism by up to 20 percent among gang members.[31]

Pittsburgh, Pennsylvania: Predicting Violence and Homicide Among Young Men

Since 1999, homicide has remained the second leading cause of death for U.S. residents 15–24 years of age. In 2010, 4,828 people in this age group were victims of homicide.[32] Like other urban communities, the Pittsburgh region has suffered lasting trends in violence among boys from late childhood to early adulthood. The ability to predict

---

[29] Jean-Pierre Guay, *Predicting Recidivism with Street Gang Members*, Ottawa, Ont.: Public Safety Canada, 2012.

[30] For more information on the LS/CMI, see Multi-Health Systems, Inc., "LS/CMI: Product Overview," web page, undated.

[31] James Bonta and D. A. Andrews, *Risk-Need-Responsivity Model for Offender Assessment and Rehabilitation*, Ottawa, Ont.: Public Safety Canada, 2007.

[32] National Center for Injury Prevention and Control, Centers for Disease Control and Prevention, "Youth Violence Datasheet," Atlanta, Ga., 2012. This case study draws on Rolf Loeber, Dustin Pardini, D. Lynn Homish, Evelyn H. Wei, Anne M. Crawford, David P. Farrington, Magda Stouthamer-Loeber, Judith Creemers, Steven A. Koehler, and Richard Rosenfeld, "The Prediction of Violence and Homicide in Young Men," *Journal of Consulting and Clinical Psychology*, Vol. 73, No. 6, 2005.

homicide in populations of boys would provide law enforcement and youth-focused community organizations with the resources to identify boys at particularly high risk and intervene.

The researchers considered as predictors of violence a class of risk factors that included child, family, school, and demographic characteristics. The research team sampled 1,517 total boys from three schools in the Pittsburgh region (including 33 convicted of homicide, 193 convicted of serious violence, and another 498 self-reporting serious violence) using a composite behavioral instrument with 63 risk factors based on prior literature. The survey results were then used to construct a violence risk index (based on 11 of the original factors) and a homicide risk index (based on nine factors). The research provides evidence that it is possible to estimate the risk of violence in a community sample of boys.[33]

Risk Assessment for Organized Crime Behavior

Predicting crimes committed by organized gangs differs from detecting crimes committed by individuals and predicting who will offend. Criminal gangs usually engage in one or two types of criminal behavior, such as drug-related crimes or prostitution. In addition, criminal gangs often fight each other to gain dominance in certain parts of the city. However, our review of predictive methods would not be complete without some mention of attempts to forecast these types of crimes.

Predicting events related to organized crime was discussed very briefly at the October 2009 NIJ Geospatial Technology Working Group meeting in Scottsdale, Arizona. The discussions centered on the idea that technically savvy criminals can be highly adaptive and tend to look for opportunities to employ new technology in their criminal acts. For this reason, police and analysts must continually educate themselves on these innovations.

The example of human trafficking enabled by some virtual spaces, such as Craigslist and other open and less regulated online discussion forums, has been raised as a type of criminal behavior enabled through technology. Meeting participants pointed to organized crime groups as behind these more industrious crimes, stating that technology has broadened the reach of these groups and made their activities less geographically centralized than ever before. They also noted that policing financial and human trafficking cybercrimes requires a major change in current practices.[34]

The mobility of criminal groups is of international interest because crime organizations can influence both local crime (e.g., through illicit drugs, prostitution, and rob-

---

[33] Loeber et al., 2005.

[34] Ronald E. Wilson, Susan C. Smith, John D. Markovic, and James L. LeBeau, *Geospatial Technology Working Group Meeting Report on Predictive Policing*, Scottsdale, Ariz., October 2009.

bery) and, in some cases, international crime (e.g., through human trafficking, money laundering, and the transport of illicit drugs). Methods for predicting the illegal activity of organized crime groups include using the activity and dynamics of the group to evaluate risks and techniques to measure criminal market opportunities. The most common models include using group activity to identify organized crime,[35] determining the presence of organized crime through illicit markets,[36] and using a risk-based assessment (a more holistic approach) to identify illicit behavior.[37] For the most part, to date, these methods have been qualitative rather than formal statistical models for calculating the risk of "organized crime."

Motivating the more holistic risk-based assessments of illicit behavior are questions like "Is there much organized crime?" "Is the situation serious?" "Is it bad that there are more criminal groups now than in the past?" and "Which criminal groups are the most dangerous?" Unlike the behavioral assessments used by the correctional system, a method developed by Tom Vander Beken, working for the Belgian Federal Police and Minister for Justice, measures organized crime starting with a general framework for organized crime. This type of crime, the study said, is entrepreneurial, and fluctuations in the market space explain the behavior of criminal groups. The methodology incorporates three major components: environmental scanning, analysis of criminal organizations and counterstrategies, and an analysis of licit and illicit market performance. Similar qualitative methods have been used to forecast crime. The following list identifies and defines several of the methods used by law enforcement agencies to describe trends in the criminal market space:

- *Environmental scanning:* A systematic effort to identify future developments that could plausibly occur and whose occurrence could alter a particular environment in an important way (e.g., economic fluctuations, social attitudes, advances in technology). Reviewing and synthesizing the literature in disciplines relevant to the issues at hand are the two most common methods used by practitioners.
- *Nominal group and Delphi process:* Surveying the opinions and judgments of experts can also provide the insight needed to assess the potential of various future events to alter the characteristics of the criminal market space. The Delphi process is one such expert survey protocol in which a group of experts are asked to debate the probability of an event and, through the process, are encouraged to find consensus.

---

[35] Mohammad A. Tayebi, Uwe Glässer, and Patricia L. Brantingham, *Organized Crime Detection in Co-offending Networks*, working paper, Burnaby, B.C.: Simon Fraser University, c. 2011.

[36] Jay S. Albanese, "Risk Assessment in Organized Crime: Developing a Market and Product-Based Model to Determine Threat Levels," *Journal of Contemporary Criminal Justice*, Vol. 24, No. 3, August 2008.

[37] Tom Vander Beken, "Risky Business: A Risk-Based Methodology to Measure Organized Crime," *Crime, Law, and Social Change*, Vol. 44, No. 5, June 2004.

- *Scenario writing:* An attempt to identify the range of possible conditions, scenario writing challenges an analyst team to use facts, given events, and a series of proposed forces to develop a picture of the conditions that may lead to various potential outcomes. By varying the assumptions associated with existing trends and considering the dynamics of the environment, analysts attempt to prioritize key actors, conditions, and events to watch.[38]

The operations that employ these techniques for studying organized crime and the market space for criminal activity are generally within the purview of international agencies charged with monitoring and deterring large-scale criminal activities. It is still useful for local police departments to understand these techniques and how they have been applied, however. In general, the methods are designed more to identify opportunities for crime than to monitor the nature of the group itself. Reviewing a range of international crime prevention activities, Morselli, Turcotte, and Tenti focused on the economic push and pull factors that drive criminal interest toward one location over another. The authors propose several tactics, as part of their discussion of criminal market spaces, to restrict the influence of organized crime.[39]

As a means to help policymakers set strategic priorities, several national and international governments have developed organized crime threat assessment mechanisms. These monitoring systems, tests, and protocols provide a framework for filtering and analyzing data about the actions of known crime groups and the crimes with which they are potentially associated. The systems vary in reliability and validity, however. Andries Johannes Zoutendijk presents a series of case studies of the major European threat assessment systems to illustrate these challenges.[40]

An important finding from the extensive review in Zoutendijk's article is that the concepts of organized crime, threat, and risk are ambiguous, and this ambiguity creates challenges for assessing the validity of one threat assessment system over another. Proper measurement instruments are still in development nationally and internationally, and these innovations may provide opportunities to improve the tools best suited for other law enforcement agencies. Natasha Tusikov shares several useful case studies of policing methods that use risk assessment methods to analyze organized crime. Like Zoutendijk, Tusikov concluded that there is ambiguity in the definitions and analysis

---

[38] Stephen Schneider, *Predicting Crime: The Review of Research*, Ottawa, Ont.: Canadian Department of Justice, 2002.

[39] Carlo Morselli, Mathilde Turcotte, and Valentian Tenti, *The Mobility of Criminal Groups*, Ottawa, Ont.: Public Safety Canada, 2010.

[40] Andries Johannes Zoutendijk, "Organized Crime Threat Assessments: A Critical Review," *Crime, Law, and Social Change*, No. 54, No. 1, August 2010.

procedures adopted by police agencies to assess risk in the case of organized crime and that this is a serious weakness.[41]

As mentioned, these measures are largely qualitative, but there has been some movement toward quantitative modeling. In 2012, for example, the Chicago Police Department experimented with social network analysis methods developed by sociologist Andrew Papachristos. These methods showed that an individual co-arrested with a homicide victim—or who is a few social links away from a homicide victim—may be orders of magnitude more likely to be killed.[42] How to take action on information like this is very much an open question.

Risk Assessment Instruments for Domestic Violence

In Johnson County, Kansas, members of the district attorney's office developed a "lethality assessment" that must be completed by law enforcement officers on the scene of domestic incidents where there is probable cause for arrest. Several risk factors, the office determined, are associated with an increased risk of homicides of women and men in violent relationships. While these associations are difficult to predict with precision, the district attorney's office was aware that higher risk factors come with a higher likelihood of homicide. Using the results of the lethality assessment administered to the victim, officers were able to automatically trigger certain protocols, including a call or referral to the SAFEHOME domestic violence shelter hotline. In 2011, the SAFEHOME shelter received around 1,000 more calls to its hotline than in prior years (an approximately 25-percent increase).[43]

Nearly 60 police departments in Maryland are using a similar lethality assessment checklist to identify potential domestic violence threats. Jacquelyn Campbell, a nursing professor at Johns Hopkins University in Baltimore, developed the questions used by Maryland's intervention. The protocol was originally designed for abuse-victim advocates and health professionals. The Maryland program charges police officers and call operators to intervene as part of domestic calls, asking the apparent victim (generally a woman) a series of pointed questions. Depending on the answers provided, the offices involved may immediately call a domestic violence counselor to guide

---

[41] Natasha Tusikov, "Measuring Organised Crime–Related Harms: Exploring Five Policing Methods," *Crime, Law, and Social Change*, Vol. 57, No. 1, February 2012.

[42] For example, if A is co-arrested with B, who was co-arrested with homicide victim C, then A is at a very high risk of being a victim as well, making A very high-risk. Andrew V. Papachristos, Anthony A. Braga, and David M. Hureau, "Social Networks and the Risk of Gunshot Injury," *Journal of Urban Health*, Vol. 89, No. 6, December 2012; David Kennedy, "After a Horrific Summer of Murder, Chicago Trying a Bold New Approach," *The Daily Beast*, September 28, 2012.

[43] Tess Koppelman, "Lethality Assessment Leads Women to Domestic Violence Shelter," WDAF-TV, Fox 4 News, Kansas City, July 23, 2012.

the victim in taking positive steps to protect him- or herself. In an interview with a *Baltimore Sun* reporter, Corporal Tracy Farmer of Maryland's Harford County Sheriff's Office remarked, "As first-responders, we're getting there in the heat of the moment. . . . If you get these victims a couple of days later, their batterer will be trying to make amends and the victims will have had time to rationalize [the assault]. It's helpful not only to tell them of the resources available, but to get the ball rolling."[44] Three primary questions are calculated to reveal direct threats of deadly violence:

1. Has your partner [or whoever the aggressor is] ever used a weapon against you or threatened you with a weapon?
2. Has he or she ever threatened to kill you or your children?
3. Do you think that he/she might try to kill you?[45]

Different from the more structured behavioral instruments used by correctional systems and those used to identify violence among populations of young boys, the lethality assessment is a semistructured interview that allows the officer greater leeway in assessing the domestic condition and the tools to take informed action quickly. Variation in the level of "control" an interviewer has over the interaction determines the interview approach (i.e., the less control results in a less structured interview, and the more control means more structure). With less structured interview protocol, an interviewer exerts limited control over the course of the discussion. Although the responses may be rich, they may also be more difficult to assess and compare to other standards. Some expert interviews to determine criminal market spaces and forecast crime trends may be conducted with an unstructured interview protocol. A semistructured interview will incorporate a guide for the questions and topics, as well as an established method to evaluate the potential responses. Highly structured interview protocols, like the behavioral instruments described earlier in this chapter, are much more rigid in their design but may be more statistically reliable and valid.[46]

According to Campbell's research, women who have been threatened with a gun are 20 times more likely to be murdered, and women who have been threatened with murder are 15 times more likely to be killed. Additionally, nearly a third of the victims who speak to a counselor eventually seek a protective order, shelter, counseling, a support group, or other services. Since 2005, the Maryland Network Against Domes-

---

[44] Justin Fenton, "Tool Gauges Abuse Risk: Program Assesses Danger in Cases of Domestic Violence," *Baltimore Sun*, November 14, 2007.

[45] Charles Remsberg, "Lethality Assessment Helps Gauge Danger from Domestic Disputes," *PoliceOne.com*, December 12, 2007.

[46] Margaret C. Harrell and Melissa A. Bradley, *Data Collection Methods: Semi-Structured Interviews and Focus Groups*, Santa Monica, Calif.: RAND Corporation, TR-718-USG, 2009.

tic Violence has developed training materials and other resources around Campbell's lethality assessment and continues to lead the program for the state.[47]

Risk Assessment Instruments for Mental Health

As the United States has moved toward integrating the mentally and physically challenged into mainstream society over the past several decades, law enforcement has had to prepare for interactions and conflicts with these individuals. Often, law enforcement officers are called upon to assess not only a general risk of violence but also specific types of violence, sometimes with only moments or even seconds to make those decisions. Now more than ever, agencies are pairing up with mental health professionals to make these assessments well in advance, or at least reasonably in advance of critical situations. In Olathe, Kansas, Johnson County Mental Health has paired with the police department to provide a full-time mental health professional dedicated to responding to calls involving mental health patients or with obvious mental health problems. Not only is the mental health professional fully trained for such encounters, she is able to provide immediate insight and information to officers on the scene to protect the patient, the officers, and other citizens. This model proved to be effective during the pilot implementation, and other nearby jurisdictions are now seeking funding to implement similar programs.

However, reliance on behavioral patterns is extremely problematic in that there is no agreed-upon list of behaviors that point to likely offenders with any certainty. Further, individuals with mental health disorders can exhibit threatening behaviors but are not necessarily potential offenders. Determining which ones are truly likely to commit serious crimes is far from a settled science.

Predictive Methods: Finding Suspects

In this section, we review several methods currently in use to "predict" who likely committed recent crimes of interest. The common approach is to assemble the available clues—pieces of data both about the crime and about past perpetrators in the area—and, using a combination of matches to potential suspects and exclusions based on the process of elimination, identify the most likely perpetrators. In all of these methods, the issues of privacy and civil rights loom large. Where possible, we report on the success or failure of the methods highlighted here.

---

[47] Maryland Network Against Domestic Violence, "What Is LAP?" web page, undated.

Basic Queries

The use of various queries of intelligence and master name databases are useful in finding suspects. Information collected through database queries of persons under supervision, field interview cards, and gang intelligence, for instance, assists investigators and analysts in identifying threats and finding likely suspects.

Criminal Intelligence in Social Network Analysis Format

Clearly, social network analysis has become extremely popular and important in recent years. Monitoring real-time updates on selected Facebook, Twitter, and other social media can provide law enforcement officials with immediate information on crimes just committed and other criminal activity being planned. From criminal mischief by juveniles to very serious crimes, including homicide, some criminals frequently post and promote their criminal activity and plans, leaving the ball in law enforcement's court to identify and stop it, when possible.

Links to Department of Motor Vehicle Registries, Pawn Data, and Other Registries

Classified as third-party data sharing agreements, law enforcement partnerships with motor vehicle and other public registries have been used to develop rich profiles of potential suspects, known offenders, and possible witnesses. One data-sharing network, in Minneapolis, is a well-known case that provides valuable evidence on the effectiveness of regional and cross-jurisdiction partnerships. Like many cities, Minneapolis experienced mortgage-default problems on a wide scale, which has led to foreclosed properties, vacant homes, and a gradual increase in crime (specifically, copper theft). The Vacant House Project Coalition was established to integrate data sharing across public service agencies throughout the city and across community associations to produce neighborhood-level data.[48] Pawn shop data have also proved to be very useful. In addition to identifying attempts to pawn known stolen property, a review of pawn records can show an increase in pawns by a particular subject that may warrant further investigation. An analysis of those data, coupled with information and intelligence from other sources, such as the areas frequented by the subject and the types of crimes committed, can inform behavioral and appropriate responses to target the person or the activity.

Anchor Point Analysis, or Geographic Profiling

Geographic profiling is an analytic tool that determines the most probable area of an offender's search base through an analysis of his or her crime locations. For the vast majority of criminals, their search base is their residence. In some cases, the search base for an offender's crimes is some other anchor point, such as his or her work site

---

[48] Jacob Wascalus, Jeff Matson, and Michael Grover, *Assembly and Uses of a Data-Sharing Network in Minneapolis*, Washington, D.C.: Board of Governors of the Federal Reserve System, last updated April 4, 2012.

or immediate past residence. Either type of base can be located using this technique. Several major software programs are being employed by police agencies to access these capabilities, including Rigel, CrimeStat III, and Dragnet.[49]

Geographic profiling is typically applied to a crime series, a set of crimes believed to have been committed by the same offender. Each crime location gives information about the offender's awareness space; more crime locations provide more information about the offender and produce a better geographic profile. Some predictive tools predict "outward" by attempting to forecast the next location to be hit by a serial criminal; geographic profiling focuses "inward," attempting to predict the offender's residence.

While these various software programs use different types of distance-decay functions to model the geographic space associated with a spree of crimes, the general output is essentially the same. The applications produce a grid over an area and then calculate the probability that the offender's home base of operations is in each grid cell based on the spatial relation to the crimes. Some software systems display this grid using a three-dimensional diagram, where X and Y correspond to map coordinates, and the vertical height reflects the probability estimated for each cell. Law enforcement officials have used these tools to prioritize suspects and tips, for designing patrol and surveillance strategies, and for other area-focused investigation tactics to narrow in on an offender.[50]

Figure 4.1 is a screenshot of the Rigel geographic profiling software from Environmental Criminology Research, Inc. Rigel can display crime maps as well as the output from a geographic profiling analysis.

Geographic profiling works by making inferences about the spatial characteristics of an offender. Computerized geographic profiling software assists in the production of color-coded maps that show the most likely area of the offender's search base. Investigators use the geographic profiling maps to focus on the best locations to find the offender with the goal of locating him or her sooner rather than later.

It is important to note that one does not necessarily need specialized computer software to do a basic geospatial profiling analysis. Analysts can do reasonably well identifying areas containing anchor points just using the heuristic that offenders tend to have residences (or other anchor points) near where they commit crimes.

Modus Operandi Similarity Analysis

Clearly, examining individual offenders based on their known modus operandi (MO) to crimes that have been committed creates reasonable leads for investigators to follow

---

[49] National Institute of Justice, "Geographic Profiling," web page, last updated December 15, 2009. Rigel is a geographic profiling application developed by Environmental Criminology Research, Inc. Dragnet was developed by David Canter at the Centre for Investigative Psychology, housed at the University of Huddersfield in the UK.

[50] Tom Rich and Michael Shively, *A Methodology for Evaluating Geographic Profiling Software*, draft report, Cambridge, Mass.: Abt Associates, prepared for the U.S. Department of Justice, December 2004.

Figure 4.1
Rigel Geographic Profiling Tool

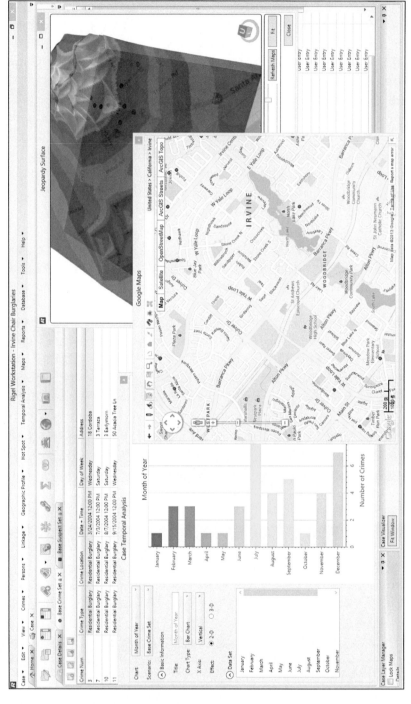

SOURCE: Courtesy of Environmental Criminology Research, Inc. Used with permission.
RAND RR233-4.1

when following up on past incidents. However, using this same information on offenders, analysts can project future behavior. For instance, suppose a known burglar targeted aging strip malls in a given county where he found an unsecured door or window for entry and then cut drywall to gain access to adjacent businesses. This type of case can be examined on a microscopic level to predict future activity. By asking such questions as "Has this offender ever repeated any targets?" and "What is the age range of the strip malls he has targeted?"—along with examining all possible targets in the jurisdiction and conducting temporal analysis—analysts can perform next-incident predictions on an individual basis.

This sort of analysis can be done manually or with computer assistance. In the manual version, the analyst sets up a table that compares key attributes of crimes committed by a known offender to other crimes that have not yet been matched; a crime that matches a large fraction of the attributes is probably part of the same crime series committed by the same perpetrator. In the automated version, computer software computes a probability that a recent crime is part of a crime series. For example, Rigel, described earlier, has a module that formally estimates the probability that recent crimes are part of an established crime series.

Although this type of analysis is quite common in many local agencies, especially those with progressive analysts, responses to this information are often lackluster. Police agencies are unsure what to do with a forecast that "John Smith, a known drywall burglar, is likely to strike again this month in one of six possible targets, based on his past behavior, MO, and target selection." While interesting, a lack of resources to develop practical and cost-efficient responses prevents most agencies from using their analysis in a proactive way.

One such real-life example from Shawnee, Kansas, involves a prediction made by co-author Susan Smith when she was the local department's crime analyst. Smith examined the future behavior of a small group of individuals who were stealing from change machines that convert dollar bills into quarters. The suspects, a set of common-law spouses, a brother, and a brother-in-law worked in concert to repeatedly target several car washes at night, using a crow bar to pry open the coin-boxes and remove the cash. Based on the individual MO factors of the primary suspect, Smith developed and disseminated a special bulletin on the particular behaviors associated with the incidents that had occurred in the jurisdiction. The bulletin included a prediction of similar activity occurring at one of three possible locations that weekend, in the hours just after dusk. Although no special personnel were assigned to the forecast, one particularly proactive officer printed the bulletin and carried it with him while on patrol. He pulled into the parking lot of one of the named targets, positioning his car out of view but in a location where he could observe the car wash while working on a report. The officer had been in place less than ten minutes when the suspects drove up and parked in a car wash bay. After three subjects were subsequently taken into custody, the primary suspect provided a false name to the arresting officer. The primary officer on the call pulled the folded bulletin from his pocket and showed the suspect a picture of himself, along with his real name, home address, associates, vehicles, and MO information. The suspect was also charged with providing false information. In this case, knowing the subjects and their MO and using this information to predict individual future behavior proved effective.

Exploitation of Sensor Data

Increasing volumes of sensor data provide growing but largely unexploited opportunities to find perpetrators. Analysts could, for example, determine whether the same license plate was spotted repeatedly near multiple burglaries or robberies that appear to be part of the same crime series. Similarly, jurisdictions have increasingly turned to GPS offender-tracking devices (such as ankle bracelets) to assist in supervising those on probation or parole. Analysts can perform queries of GPS tracking databases to see whether any tracked offenders were in the immediate vicinity of a crime. Facial recognition software, as it improves, will add yet another potential source to identify perpetrators who commit crimes in front of surveillance cameras.

Putting the Clues Together: Multisystem and Network Queries

Fusion and crime analysis centers have continued to leverage federated searching technologies to integrate disparate databases into a standardized query and analysis platform. The ability to conduct "one-stop" searching across the data sets of multiple jurisdictions and agencies has substantially enhanced the ability to translate investigative leads (such as aliases or phone numbers) into known suspects, identify multijurisdictional crime patterns, and discover offender associations and criminal networks. Data integration on such a large scale, combined with query and analytic tools to exploit the data, allows departments to leverage all available clues about crimes and known potential perpetrators to zero in on the most likely suspects. This can be a very powerful tool, with the computer able to search across huge volumes of data far exceeding what any one analyst or detective can know.

While such technologies currently supplement more traditional workflows, the integrated nature of federated data systems has great potential to enhance predictive policing efforts. At the simplest level, federated systems allow existing predictive algorithms to be applied across jurisdictional boundaries, making multijurisdictional predictive policing efforts possible. At the more sophisticated level, offender risk and suspect detection algorithms could be integrated into a federated data environment and combined with alerting functionality to automatically notify officers when they encounter a high-risk offender from outside their jurisdiction. When building predictive algorithms, federated systems also have the benefit of providing the researcher with broader and richer source data with which to develop and test models. As of this writing, no major predictive policing efforts have applied predictive technologies or methods to federated systems, but the nature of such systems is ripe for integration with predictive techniques, and such integration has the potential to substantially improve analytical capacity. The following are just a few examples showing the potential power of federated data analysis:

- The Vancouver Police Department was faced with a shooting victim who was a known gang member, with the only clue to the shooters being a partial description of the vehicle used in the incident. They used queries of gang intelligence data on known adversaries of the gang member, combined with vehicle registry data, to zero in on suspects matching the intelligence information and whose car matched the vehicle description.[51]
- The Baltimore Police Department was faced with a homicide victim who was also a known gang member, with the only clue being a surveillance video that showed the likely perpetrators getting on a train from a distance. Officers used queries of gang intelligence on the known adversaries of the gang member, combined with the knowledge that the perpetrators were likely members of a rival gang based in the direction of the train, to rapidly narrow down the potential suspects.[52]
- A tool that IBM is developing for the Miami Police Department combines data on the physical characteristics of a robbery suspect with data on prior offenders who live or work nearby. The tool is intended to help law enforcement agencies create a rank-ordered list of the most likely suspects. It is reportedly highly effective in identifying the suspects, assuming they have been prosecuted for prior robberies.

Taking Action on Predictions

The first stage in taking action on a prediction about a high-risk individual or potential suspect is to describe the crime pattern or evidence related to the person.[53] Past this point, actions taken with regard to individuals become far more uncertain and of greater concern from a privacy and civil rights perspective. This is an area that calls for significantly more research. However, in this section, we review practices that are currently in use.

---

[51] Jason Cheung and Ryan Prox, Vancouver Police Department, "Fighting Crime with CRIME: A Single, Integrated Analytical, and Investigative Umbrella," presentation to the International Association of Chiefs of Police Law Enforcement Information Management Conference, Indianapolis, Ind., May 23, 2012.

[52] Kerry Hayes, "Baltimore Police Department: Incorporating Technology to Reduce Violence," presentation to the International Association of Chiefs of Police Law Enforcement Information Management Conference, Indianapolis, Ind., May 22, 2012.

[53] International Association of Crime Analysts, *Crime Pattern Definitions for Tactical Analysis*, White Paper 2011-01, Overland Park, Kan., 2011.

Identifying High-Risk Individuals

We found evidence that police departments and correctional systems have used following interventions, among others, in response to high-risk individuals:

- One basic measure is to support improved situational awareness of the highest-risk offenders by providing officers with updated priority lists of known offenders.
- For especially high-risk individuals, some departments allocate resources to conduct regular surveillance.
- For high-risk individuals under community supervision (probation and parole), departments and corrections agencies have applied increased supervision, such as additional meetings and checks or the use of GPS tracking devices.
- Interventions led by community or other nonprofit organizations are also in use. For example, Cure Violence (formerly Chicago Ceasefire) is widely known for intervening with individuals at high risk of involvement in gang-related or other violent attacks.[54]
- Some intervention and therapy programs have shown promise for treating persons at high risk of engaging in domestic violence, mental health–related violence, or violent behavior in general. CrimeSolutions.gov identifies a number of programs with demonstrated effects in reducing incidents of violence.[55]

Identifying the Most Likely Suspects

Actions taken in response to predictions of the most likely suspects are more straightforward. Here, the actions involve simply integrating the findings into the criminal investigation process. The complications relate to due process and civil rights: What levels of statistical "risk" or "certainty" constitute standards for taking various degrees of actions against suspects? This is likely to be a major topic of future research and debate.

Prediction-Based Offender Intervention in Context

The following case studies profile the application of a few of the techniques discussed earlier. The emphasis in these cases is on predicting who will commit a crime and why. The sources of these examples are the various departments that implemented the prediction methods. As with our case studies of predicting where and when crimes will occur, there has been no independent verification of the claims made concerning the

---

[54] See Cure Violence, homepage, undated, and "The Interrupters," television broadcast, *PBS Frontline*, February 14, 2012.

[55] U.S. Department of Justice, Office of Justice Programs, "Crime and Crime Prevention: Violent Crime," web page, undated.

success of these methods. We include them here to illustrate how various departments are using predictive methods to fight crime in their communities.

Florida State Department of Juvenile Justice: Preemptive Efforts

Despite a decline in the national population of incarcerated adults, several state prison systems are overcrowded and growing.[56] In a 5-4 decision on May 23, 2011, the U.S. Supreme Court found prisons in California to be overcrowded and ordered the state to reduce the size of its inmate population by more than 30,000.[57] California's prison population had already been declining and will decline further as a result of the decision. Florida, in contrast, saw steady growth in its prison population between 2008 and 2009. The state's prison population grew in absolute size by 1,527 inmates (a 1.5-percent increase) during that period, the second largest state-level increase. A research study by the Pew Center on the States captured these trends and offered several reasons for why the national decline in 2009 might be the start of a longer-term decline in the national prison population. One of the proposed reasons is that better use of data and analytic tools to predict the potential for additional criminal behavior can improve a case management system's ability to cut the chance of reoffending.[58]

Part of the increase in Florida's inmate population might be attributable to repeat juvenile offenders who have a high probability of serving time in prison as adults. The Florida Department of Juvenile Justice's Bureau of Research and Planning manages the Juvenile Justice Information System (JJIS), one of the largest and most comprehensive state-operated databases of its kind. Encompassing the criminal history of more than 1 million youth back to the year 2000, the Department of Juvenile Justice uses the JJIS data to design juvenile intervention programs and craft department-level strategies. Specifically, the department's statistical profiling program uses the large historical data set to predict the successful placement of juveniles in rehabilitation programs. Although this is not a traditional predictive policing technique used to craft specific "actions" for forecasted problem neighborhoods, the effort keeps recidivism rates down and neighborhoods safer as a result.

In 2007, the department joined with IBM SPSS to pilot a statistical profiling program to predict the juvenile criminal behaviors of 85,000 youth and assign them to predicted risk-specific rehabilitation programs in an effort to reduce recidivism. Youth were assigned to programs based on the department's experience assigning other, simi-

---

[56] Pew Center on the States, *Prison Count 2010: State Population Declines for the First Time in 38 Years*, Washington, D.C., April 1, 2010.

[57] Adam Liptak, "Justices, 5-4, Tell California to Cut Prison Population," *New York Times*, May 23, 2011. In the Supreme Court's opinion in *Brown v. Plata*, Docket No. 09-1233, 2011, Justice Anthony Kennedy emphasized that the reduction in California's prison population need not be achieved solely by releasing prisoners early. He suggested that new prison construction, transfers out of state, and using county facilities to manage the population would be sufficient.

[58] Pew Center on the States, 2010.

lar, youth to programs in the past. Some programs aligned youth with sponsor families, community members, and educational opportunities, while others were more oriented toward discipline and skill development. By tailoring the assignment process to link youth to the optimal program by statistical analysis the state hoped to reduce recidivism overall.

Although the program has not been scientifically tested, the state reported success: Six months after the preventative services assigned to the youth during the pilot had ended, 93 percent of the 85,000 youth remained arrest-free. "We are hoping that the use of predictive analytics will essentially put us out of business, with first-offending juveniles never returning to the incarceration system," said Mark Greenwald of the Florida Department of Juvenile Justice in an article about the pilot program.[59]

As of fiscal year (FY) 2009–2010, the department had seen statewide delinquency referrals decline by 15 percent; school referrals were down by 9 percent, and the number of juvenile offenders waiting in detention centers for placement was at a historic low. The program, renamed the Residential Positive Achievement Change Tool (R-PACT), was launched statewide in FY 2009–2010 after its success in FYs 2007–2008 and 2008–2009.[60] R-PACT and a similar probation-oriented program (referred to as Community-PACT) have helped the state identify at-risk youth and assign them to low-risk, moderate-risk, high-risk, and maximum-risk rehabilitation programs.[61] By using data to predict the behavior of youth offenders, Florida aims to reduce its adult prison population over the long term.

Predicting Predator Hunting Patterns

The hunting pattern of the male lion, when aggregated to a time-series and geospatial research display, has been described as looking like a daisy. The "petals" of the lion's floral-like hunting pattern trace its search for prey and the time taken to carefully stalk, kill, and then return home to feast. Although predicting the next kill point in the lion's behavioral pattern is very difficult, some criminologists suggest that estimating the center of activity—the hunter's rest site—may be less so.[62]

According to Kim Rossmo, serial killers, rapists, and arsonists, among other serious offenders, share behavioral characteristics that are innate to all hunters. Rossmo has made a career of using geographic profiling techniques to assist in ongoing inves-

---

[59] Mark Greenwald, "Improving Juvenile Justice for the State of Florida," *Building a Smarter Planet: A Smarter Planet Blog*, April 14, 2010.

[60] In FY 2007–2008, the predictive analytic information provided through JJIS (now R-PACT) used evidence based methods to improve the assignment of youth to Department of Juvenile Justice rehabilitation programs, which helped keep recidivism at 43 percent. The move to expand the program statewide is projected to further reduce the recidivism rate.

[61] Florida Department of Juvenile Justice, *Fiscal Year 2009–10 Annual Report*, Tallahassee, Fla., 2010.

[62] Bruce Grierson, "The Hound of the Data Points," *Popular Science*, March 21, 2003.

tigations as a Texas State University research professor and, previously, as a detective inspector with the Vancouver Police Department in Canada. Rossmo coined the term *geographic profiling* while pursuing doctoral research at Simon Fraser University's School of Criminology in British Columbia, and he founded a section of the Vancouver Police Department focused on these crime analysis methods. The concept of using geospatial and other tactical information to construct behavioral patterns has been tested with evidence used to convict criminals, identify the origin of disease outbreaks, and find the resting sites of bees, mosquitoes, and great white sharks. In his doctoral dissertation, Rossmo traced the predation patterns of psychopathic serial killers, such as Richard Trenton Chase, the "Vampire Killer," to their resting areas to within 1.7 percent of the total hunting area. As Rossmo has indicated, geographic profiling can filter the tremendous amount of investigation data points to indicate where the hunter rests.[63]

The process was instrumental in helping police in Lafayette, Louisiana, identify the South Side Rapist, who had successfully evaded the police for almost a decade. Unidentified, the assailant had stalked the streets of Lafayette, concealing his face from identification with a wrapped scarf. Thousands of tips and hints about the case had pointed to thousands of suspects; the data scale was too great for the Lafayette Police Department to analyze in full. Rossmo was hired in 1998 to help sort through the information with his geographic profiling techniques. Rossmo's deductive analysis helped police filter through the overwhelming number of suspects with a simple assessment of where each resided in relation to the hunter's predicted resting site.[64]

Although the first pass through the data and the suspect list did not uncover the killer, the process changed the police department's thinking about managing the suspect list. When a new suspect was named, rather than appending the name to the growing list or disregarding it on face value, police checked it against the geospatial information Rossmo had developed. The new suspect—a sheriff's deputy from a neighboring police department, Randy Comeaux—had lived in the predicted area during the rapes. The police put a tail on Comeaux, collected one of his tossed cigarette butts, and used it to test his DNA against the killer's. The DNA matched. Without Rossmo's attention to geospatial data, Comeaux may have maintained his double life—officer of the law by day, South Side Rapist by night.

The algorithm Rossmo developed is based on two concepts crafted by his graduate school advisers at Simon Fraser University. Paul and Patricia Brantingham developed the conceptual frameworks for thinking about an offender's "buffer zone" and the "distance decay" that describes the way an offender ventures out from his or her resting site to commit a crime. The "buffer zone" suggests that a hunter will hunt near

---

[63] D. Kim Rossmo, *Geographic Profiling: Target Patterns of Serial Murders*, doctoral thesis, Burnaby, B.C.: Simon Fraser University, 1995.

[64] Grierson, 2003.

his or her home area, but there will be a buffer of some distance around the resting site to maintain a safe distance from the site of the crime. Rossmo mechanized both of these concepts into his algorithm, which he has since patented. Several similar algorithms have also been developed, including CrimeStat, Dragnet, and a circle and range model created by Canter and Larkin.[65]

Geographic profiling evaluates the probability that each cell in a grid across a specific geographic region will contain the resting site of the hunter. The probabilities are based on evidence of the hunter's behavior and use the "buffer zone" and "distance decay" as methods to refine the scope of the hot spots. These maps generally look like topographical maps of canyon regions with low-probability sites at the bottom of deep data valleys and the highest-probability points appearing as mountain peaks. Rossmo has stressed that the data analysis provides a framework for thinking about how to catch a predator and it is not a push-button solution. "The goal is to help law enforcement, intelligence and military agencies focus their limited resources in areas that are most likely to contain what they're looking for."[66]

One reason that there is no push-button solution to this type of criminal investigation is that the hunting patterns of serial offenders differ. Geographic profiling was attempted in the investigation of the 2002 Beltway sniper attacks in the Washington, D.C., area, but the techniques were not as useful as they have been in other cases. One reason that the models were not more accurate is that John Allen Muhammad and John Lee Malvo—the men arrested in connection with the case—never established a set "home base" for long. They selected hunting grounds not in places they knew but in areas that were similar to places they knew: shopping center parking lots and gas stations. Ease of access to various forms of transportation allows serial killers to increase their mobility and establish much more complex hunting patterns. In the study of serial criminal offenders, two models are generally used to describe the offender's spatial behavior during the execution of a crime series: *Marauders* are said to operate in an area that is near their home base, whereas *commuters* are deemed to commit crimes in locations perceived as being distant from their place of residence. Advancements in transportation and mobility dramatically have enlarged the range of interest for serial offenders whose spatial behavior is explained by the commuter model, posing new challenges for geographic profilers.

Geospatial profiling emerged from criminology and geography, but applied psychologist David Canter has extensively studied its use in investigative psychology. At

[65] Ned Levine, "Crime Mapping and the Crimestat Program," *Geographical Analysis*, Vol. 38, No. 1, January 2006; David Canter, Toby Coffey, Malcolm Huntley, and Christopher Missen, "Predicting Serial Killers' Home Base Using a Decision Support System," *Journal of Quantitative Criminology*, Vol. 16, No. 4, December 2000; David Canter and Paul Larkin, "The Environmental Range of Serial Rapists," *Journal of Environmental Psychology*, Vol. 13, No. 1, March 1993.

[66] Travis Hord, "Professor Uses Math to Track Criminals, Shark Patterns," *University Star* (Texas State University), August 28, 2009.

the Centre for Investigative Psychology, housed at the University of Huddersfield in the UK, Canter has led the development of Dragnet, a software package for predicting a criminal offender's base by prioritizing surrounding regions. Very similar to Rigel and other geospatial profiling software applications, Dragnet produces a gradient map the area near an offender's predicted location so that police and other agencies can "action" the area appropriately. Dragnet also has a procedure for handling physical structures and barriers on a city street, making it more flexible than other applications.[67] Although these techniques can be used to consider individual actors, perpetual criminals, and other hunter types, Canter suggests that investigative psychology techniques are most useful for predicting criminals in high-volume crimes: robbery, arson, car theft, and burglary.[68] In this way, Canter can assist the police by leveraging the larger quantity of psychological information about people who commit high-volume crimes.

---

[67] Centre for Investigative Psychology, "Dragnet," web page, undated.

[68] John Crace, "Two Brains," *The Guardian*, November 1, 2004.

CHAPTER FIVE

Findings for Practitioners, Developers, and Policymakers

He who lives by the crystal ball soon learns to eat ground glass.

—*Edgar Fiedler*

Predictive policing is more than just a few methods for analyzing data. It is a systemic and systematic process of collecting, analyzing, and responding to data. This guide has covered many of the techniques for processing these data, along with some actions that can be taken in response to that analysis. This chapter addresses common misconceptions about predictive policing and pitfalls related to its implementation, providing suggestions for building an analytical capability. We conclude with a summary of the key lessons learned over the course of this study.

While the term label is new, many of the types of analysis that constitute predictive policing have been widely used in law enforcement and in other fields, just under different terminology. The lessons from these other fields can highlight the many well-known pitfalls that can lead practitioners astray and provide recommendations to enhance the effectiveness of these methods.

Predictive Policing Myths

As a term of art, *predictive policing* is relatively new, but it has received substantial attention. Therefore, it is important to dispel some commonly repeated myths about the field. Whether by reporters seeking to punch up headlines or vendors promoting products, predictive policing has been so hyped that the reality cannot live up to the hyperbole. There is an underlying, erroneous theme that advanced mathematical and computational power is both necessary and sufficient to reduce crime. If these methods cannot live up to this claim, there is a risk of backlash. Here, we discuss some of the most problematic myths surrounding predictive policing methods.

Myth 1: The Computer Actually Knows the Future

Some descriptions of predictive policing make it sound as if the computer knows the future. Although much of news coverage promotes the meme that predictive policing is a crystal ball, these algorithms simply predict risks.[1] The predictions made by these types of analyses are based on extrapolations from data in much the same way that an insurance adjuster extrapolates from actuarial data to assess insurance risk. These techniques are successful to the extent that crime in the future will look similar to crime in the past, in terms of geographic location or the characteristics of offenders. The computer as a tool can dramatically simplify the search for these patterns, but all of these techniques are extrapolations from the past in one way or another.

Additionally, predictions are only as good as the underlying data. Because predictive analytics is fundamentally an extrapolation from past crimes, the quality of the outcome depends on the quality of the inputs. Holes in the input data will be reflected by blind spots in the outputs. Likewise, biases in the inputs will skew the predictions. For example, the arrest of a serial offender accounting for a large number of residential burglaries in a given area would not be "known" by the software. Thus, the software will continue to forecast, or predict, the next incident (or incidents) based on what it *does* know—which is the crime pattern. A thorough process to ensure the quality and completeness of the data used in the predictions is necessary for predictive programs to be effective.

Examples of what can realistically be expected are shown in Figure 5.1. Here, we compare the percentage of crimes predicted in hot spots in the coming month (November 2009, in this case) to the percentage of populated city land in the hot spots for two types of Part 1 crimes in the Washington, D.C., metropolitan area. We compare the percentages of crimes captured in the hot spots for a regression model like those introduced in Chapter Two: hot spot mapping, perfect forecasting (i.e., knowing in advance exactly which grid cells will have the most crime—equivalent to a crystal ball), and nothing (choosing locations to focus on at random).[2]

As shown, for vehicle thefts, the regression model did noticeably better than simple hot spot mapping at capturing crimes, though the differences were not that major. However, both approaches did noticeably worse than perfect forecasting. Conversely, for thefts (not involving a vehicle), the two prediction approaches came very close to perfection, but there was virtually no difference between the regression and hot spot mapping approaches. In the latter case, the locations of thefts were very closely correlated with where they had occurred previously, whereas the locations of vehicle thefts were not nearly so related. In both cases, however, there was a major difference between using some predictive technique—even simple hot spot mapping—and doing nothing.

---

[1]   Goode, 2011.

[2]   Prior months' data were from May 2009 to October 2009.

Figure 5.1
Comparing Forecast Methods for Washington, D.C., Crimes

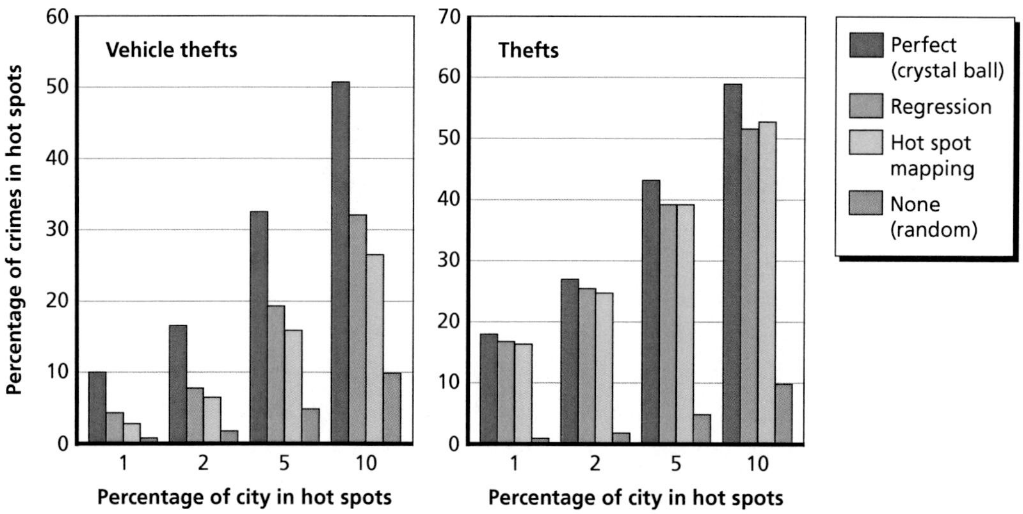

Myth 2: The Computer Will Do Everything for You

While it is common to promote software packages as end-to-end solutions for predictive policing, humans remain—by far—the most important elements in the process.[3] Even with the most complete software suites, humans must

- find relevant data
- preprocess the data so they are suitable for analysis, notably by adding identifying details and addressing any systematic data exclusions or biases
- design and conduct analyses in response to ever-changing crime conditions[4]
- review and interpret the results of these analyses to exclude erroneous findings (e.g., hot spots over water), and integrate the findings with contextual knowledge beyond the software's capabilities (e.g., knowledge that a predicted hot spot will likely disappear due to key arrests of criminals in the area)
- analyze the integrated findings in light of other demands and constraints facing the agency and make recommendations about how to act on them
- take action to exploit the findings and assess the impacts of those actions.

---

[3]   In the summer and fall of 2012, the International Association of Crime Analysts' email list saw a good bit of traffic on increasingly common questions echoing the sentiment, "What do we need analysts for when software can do it all?" We believe this finding helps answer these questions.

[4]   Except for routine statistical reporting, there will not a be standard set of analyses that one can just run all the time.

In addition to cognitive processes, these functions require a great deal of social interaction to maintain accurate contextual knowledge of the crime environment, collaborate with practitioners across the agency, and ensure that appropriate actions are taken. Computers and software can assist with many of the tasks discussed here, but they are far from able to perform these tasks themselves.

Myth 3: You Need a High-Powered (and Expensive) Model

Most police departments do not need the most expensive software packages or computers to launch a predictive policing program. While there tends to be a correlation between the complexity of a model and its predictive power, increases in predictive power have tended to show diminishing returns like that shown in Figure 5.1. As discussed in Chapters Two and Three, simple heuristics were found to be nearly as good as sophisticated analytic software in performing some tasks, such as geographic profiling and next-incident tactical predictions. As for predicting hot spots, in our research, the increase in accuracy in moving from fairly simple algorithms to the most sophisticated and computationally intensive algorithms tended to be marginal. Further, there are limits to how useful these data can be. An officer is not likely to, say, drive no more than 372 feet down a particular road to the end of a precisely predicted hot spot before turning back.

High-end software does allow for faster processing of high volumes of data, but for smaller departments with less activity, such high-end software is not needed to analyze crime data. In addition to the basic functionalities built into standard workplace software (e.g., Microsoft Office) and GIS applications (e.g., ArcGIS), there are free software tools that perform many of the same functions as the expensive software packages, such as CrimeStat III and Near Repeat Calculator, developed at Temple University with a grant from NIJ and described in Chapter Two.

Myth 4: Accurate Predictions Automatically Lead to Major Crime Reductions

Predictive policing analysis is frequently marketed as the path to the end of crime (recall the discussion of *Minority Report*). However, sometimes the focus on analysis and software can obscure the fact that predictions, on their own, are just that—predictions. Achieving actual decreases in crime requires taking action based on those predictions. Thus, we emphasize again that predictive policing is not about making predictions but about the end-to-end process.

Predictive Policing Pitfalls

To be of use to law enforcement, predictive policing methods must be applied as part of a comprehensive crime prevention strategy. And to ensure that predictive methods make a significant contribution, certain pitfalls need to be avoided. These pitfalls range from data deficiencies to assessment flaws and a failure to adequately evaluate the techniques used. In a recent article for the *National Institute of Justice Journal*, former RAND colleague Greg Ridgeway discusses seven pitfalls that he claims must be avoided if "prediction in criminal justice is to take full advantage of the strength of [predictive policing] tools."[5] The five pitfalls presented here complement his seven.

Pitfall 1: Focusing on Prediction Accuracy Instead of Tactical Utility

Suppose an analyst has been asked to provide predictions of robberies that are as "accurate" as possible. Here, "accurate" means that the analyst designs an analysis in which as many future crimes as possible fall inside areas predicted to be high-risk, thus confirming that these areas are high-risk. For example, in the case of robberies in Washington, D.C., an analyst using the regression model introduced in Chapter Two to predict robbery risk could indeed find hot spots that capture 99 percent of the risk. The result is shown in Figure 5.2.

On the positive side, Figure 5.2 probably shows where more than 99 percent of robberies will occur in Washington, D.C. On the negative side, it flags as high-risk more than two-thirds of the city area—and virtually all areas of the city with significant foot traffic. Declaring most of the city a hot spot for robbery is highly accurate but has almost no tactical utility. To ensure that predicted hot spots that are small enough to be actionable, we must accept some limits on "accuracy" as measured by the proportion of crimes in the hot spots.

To borrow an example from RAND research into counterinsurgency operations in Iraq, a computer model predicted that an IED event would occur somewhere in the city of Mosul in the next 48 hours. That was indeed accurate, but it was hardly of any tactical value. When performing tactical analysis, practitioners should generally focus on producing results with tactical utility. This means that the scale of the analysis should fit the scale of the possible responses. For example, a beat officer can likely manage a few hot spots the size of a city block but would not find it practical to focus on a strip a few miles in length.

Pitfall 2: Relying on Poor-Quality Data

Figure 5.3 shows another key pitfall in predictive policing: problems with the data used in predictions. Note that the areas of the National Mall and U.S. Capitol, as well as other landmarks (e.g., Rock Creek Park, Joint Base Anacostia-Bolling, the National Arboretum), are completely free of robbery risk. While these are indeed fairly low-risk

---

[5] Greg Ridgeway, "The Pitfalls of Prediction," *National Institute of Justice Journal*, No. 271, February 2013. Ridgeway lists the following seven pitfalls: (1) trusting expert predictions too much (eschewing technological approaches); (2) clinging to what you learned in statistics 101 (getting beyond statistics models); (3) assuming that one method works well for all problems; (4) trying to interpret too much (focus on accuracy and not transparency); (5) forsaking model simplicity for predictive strength (or vice versa); (6) expecting perfect predictions; and (7) failure to consider the unintended consequences of predictions.

Figure 5.2
Attempting to Capture All City Robberies in Hot Spots

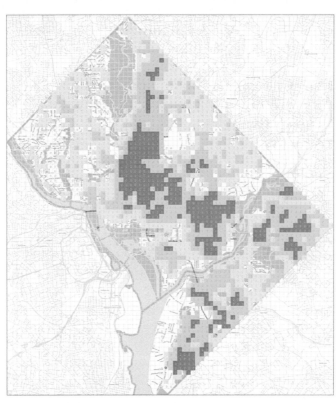

RAND *RR233-5.2*

areas, they are not no-risk. The lack of shading is due to these areas being patrolled by the National Park Police, U.S. Capitol Police, and military police, which means that the Washington, D.C., Metropolitan Police Department does not collect robbery reports for these areas. This is *data omission*, meaning that data are missing for incidents of interest in particular places (and at particular times).

Systematic errors in the data will lead to systematic errors in the resulting analysis. If data are omitted, as in Figure 5.3, it will appear that there is no crime in specific areas. This is the primary reason why it is so important for police departments to understand the ground truth in performing these analyses; department analysts can spot problems in the output that would lead to these kinds of systematic errors. A study of more than 400,000 crime incidents from six large law enforcement jurisdictions in the United States found that positional accuracy of geocoded crime events is a significant factor in predictive crime mapping. To evaluate the predictive accuracy of hot spot maps, the researchers used a series of metrics that included hit rate, PAI, and recapture

Figure 5.3
Data Omission in Robbery Reports

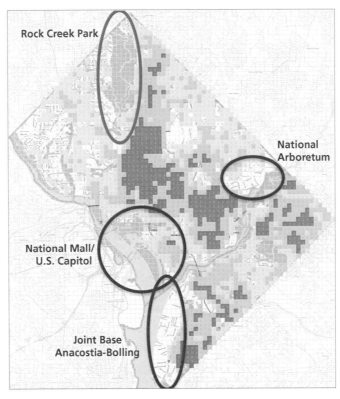

RAND RR233-5.3

rate index (RRI).[6] While the recommended data care and cleaning techniques are valuable (and outlined below), using methods like the PAI and RRI to check and validate assumptions about the data is a valuable practice for law enforcement agencies. The study's recommendations for predictive hot spot analysis include the following:

- Consider analyzing data with multiple techniques
- Disaggregate crime incidents and analyze like crime events separately
- Take study area into consideration

---

[6] Timothy C. Hart and Paul A. Zandbergen, *Effects of Data Quality on Predictive Hotspot Mapping*, submitted to the National Institute of Justice, Las Vegas, Nev., and Albuquerque, N.M.: University of Nevada, Las Vegas, and University of New Mexico, September 1, 2012. The PAI is the ratio of the density of crimes in a hot spot relative to the density of crimes in the general study area. See Chainey, Tompson, and Uhlig, 2008.

The RRI is the ratio of current-period hot spot crime density to the prior period's hot spot crime density, standardized by the total study-area density. See Ned Levine, "The 'Hottest' Part of a Hotspot: Comments on 'The Utility of Hotspot Mapping for Predicting Spatial Patterns of Crime,'" *Security Journal*, Vol. 21, No. 4, October 2008.

- Be cognizant of user-defined parameter settings
- Use street centerline reference data or address point reference data
- Determine how predictive accuracy will be measured.[7]

Likewise, it is important to understand how the data are collected because they may have systematic *biases*. If the crimes in an entire shift are reported at the end of the shift (rather than when they occur), this is just as good as 911 calls that are recorded as they occur—provided that the shift reports include the time of the incident. Figure 2.10 in Chapter Two was a heat map for burglaries by time of day and day of week. The figure showed especially heavy burglary concentrations between 7:00 and 8:00 a.m. during the workweek. However, it is not clear whether a large number of burglaries actually occurred between 7:00 and 8:00 a.m. or whether that was when property owners and managers discovered and reported burglaries that had taken place overnight.

Relevance is also an important issue with data quality. For some crime clusters, it can be very useful to have data going back many months or years. For example, if muggers frequently target bar-goers after last call, data from several months will be useful in identifying key hot spots. Conversely, if there is a spree of very similar robberies likely committed by the same criminal, several months of data will not be of much use because the data would capture both the active criminal's robberies and old cases, making patterns harder to find. In the first case, the primary commonality between the data points is the target, so it may make sense to include crimes going back as far as possible to build hot spots. (Additional information may be gleaned by looking at the evolution of the hot spots over time.) In the second case, the commonality is the likely perpetrator, and the data set used for analysis should focus on his or her crimes only.

Pitfall 3: Misunderstanding the Factors Behind the Prediction

Figures 5.3 and 5.4 highlight a third pitfall. Observers—especially practitioners tasked with making hot spots go away—may reasonably ask, "For a given hot spot, what *factors* are driving risk?" Is it prior robberies? Disorderly conduct reports? The presence of bars? The presence of recent parolees with robbery convictions? "The computer said so" is far from an adequate answer. In general, predictive tools are designed in a way that makes it difficult, if not impossible, to highlight the risk factors in specific areas. There has been some improvement, however.

When applying techniques, such as regression or any of the data mining variants, using common sense to vet the factors incorporated into the model will help avoid spurious relationships. It also is important to have a curiosity about the drivers of predictions so that deeper relationships can be found.

---

[7]   Hart and Zandbergen, 2012, pp. 61–62.

For example, testing crime occurrences against the location of police officers would likely show that there is a high correlation between where crimes occur and where police have spent time. This does not mean that we could predict the location of crimes by looking at the location of police, because the police generally show up after crimes occur. Thus, this relationship will likely be quite strong, but it provides no useful information for prediction.

Alternatively, if a strong relationship is found between certain subway stops and where crimes are occurring, this analysis could yield both a useful prediction and additional information about the criminals or their targets. It may be that these subway stops have a common link that explains where the criminals live, where their targets are coming from, and so on. Armed with this information, police could more effectively target an intervention specifically for this crime type.

This is of special concern when using data mining methods on large numbers of potential input variables: Just because the method puts a variable into a model to predict crime does not mean that that variable causes crime. Conversely, just because a variable does not show up in a particular model does not mean that the variable is not an important driver.[8]

This pitfall is closely linked to the first two myths. The human element of the analysis is key because only a trained crime analyst can provide the appropriate context to assess the utility of the computer's output. Computers may get there eventually, but they cannot perform these assessments yet.

Pitfall 4: Underemphasizing Assessment and Evaluation

During our interviews with practitioners, very few said that they had evaluated the effectiveness of the predictions they produced or the interventions that followed the prediction. As part of updating the data to keep them current, it is important to assess the effectiveness of analyses and subsequent interventions. Regardless of how well a police system operates, some aspect can almost always be improved. Measurement is key to identifying areas for improvement, determining the effectiveness of interventions, and making decisions about how resources are allocated.

To assess the effectiveness of crime analysis recommendations, follow-up data on both the crime occurrences and police responses need to be collected. The predictive models can be tested on these outcomes over time with the understanding that police responses to the predictions may have an effect on the outcomes.

On the intervention effectiveness side, for example, once a hot spot has been identified and an intervention has been employed, follow-on analysis can be used to identify the effectiveness of the intervention in that particular area. It could be that the hot

---

[8] Ridgeway's article on prediction pitfalls gives an example of multiple decision trees for predicting school dropout risks that have almost identical accuracies but use completely different input variables; the only difference was that the trees were built using different subsets of the data. See Ridgeway, 2013.

spot has been successfully suppressed for some reason, in which case police resources may be better used elsewhere. The nature of the readjustment can depend on why the hot spot has been suppressed, such as a change in the conditions or environment, an arrest of a repeat or career offender, or some other reason. Alternatively, if the hot spot persists, an alternative intervention should be considered, and additional evaluations will be needed.

Assessment and evaluation are also necessary to verify claims about software and methodologies. A vendor may claim that, following the adoption of its software, crime in a city fell by x percent. Because of broader trends in crime, this statement could have been true even if the city had not used the software. Without appropriate efforts to assess these tools, any claim of effectiveness should be taken with a healthy dose of skepticism.

Pitfall 5: Overlooking Civil and Privacy Rights

The very act of labeling areas and people as worthy of further law enforcement attention inherently raises concerns about civil liberties and privacy rights.[9] Labeling areas as "at-risk" appears to pose fewer problems because, in that case, individuals are not being directly targeted. The U.S. Supreme Court has ruled that standards for what constitutes reasonable suspicion are relaxed in "high-crime areas" (e.g., hot spots). However, what formally constitutes a "high-crime" area, and what measures may be taken in such areas under "relaxed" reasonable-suspicion rules, is an open question.[10]

Using predictive policing techniques to identify hot spots raises few privacy issues because the data typically do not contain personally identifying information.[11] Transparency about the types of information collected and the uses of that information may further help allay fears of invasion of privacy.

Taking action on hot spots poses minor challenges compared with the civil and privacy rights concerns that arise when applying similar techniques to finding "hot people." In Chapter Four, we discussed the use of various models to label individuals about to be released from detention facilities as being at risk of recidivism. What should be done with a prediction that a parolee has a high risk of reoffending when that prediction—while much better than chance—is still far from definitive? The common answer to date has been that, because most high-risk individuals are already under cor-

---

[9]   On the issue of privacy and civil rights in the context of counterterrorism operations, see Committee on Technical and Privacy Dimensions of Information for Terrorism Prevention and Other National Goals, National Research Council, *Protecting Individual Privacy in the Struggle Against Terrorists: A Framework for Program Assessment*, Washington, D.C.: National Academies Press, 2008.

[10]   Andrew G. Ferguson and Damien Bernache, "The 'High-Crime Area' Question: Requiring Verifiable and Quantifiable Evidence for Fourth Amendment Reasonable Suspicion Analysis," *American University Law Review*, Vol. 57, No. 6, August 2008.

[11]   The exception would be measures that identify specific addresses as hot spots that could be quickly tied to individuals (e.g., owners, residents).

rectional supervision of some form (or are at least convicted felons), law enforcement largely has carte blanche to take *reasonable* investigative and precautionary actions against them. Again, what "reasonable" entails, and under what conditions, is far from clear. This is an area that we believe warrants substantial further research and development over the coming years.

A Buyer's Guide to Predictive Policing

All departments can benefit from predictive policing methods and tools; the distinction is in how sophisticated (and expensive) the tools need to be. In thinking about what is needed, it is important to note that the key value in predictive policing tools is their ability to provide *situational awareness* of crime risks and the information needed to act on those risks and preempt crime. The question then becomes a matter of what set of tools best provides situational awareness for a given department, not which tools can serve as a crystal ball.

The brief summary of the tools available for small and large agencies in Chapter One provided general guidance on what might be effective for agencies with low to moderate data complexity and those with large amounts of complex data. The former types of tools tend to fall into the category of conventional crime analysis, and the latter into the more complex realm of predictive analytics.

Small agencies with relatively few crimes per year and with reasonably understandable distributions of crime (e.g., a jurisdiction with a few shopping areas that are persistent hot spots) probably only need core statistical and display capabilities. These tools are available for free or at low cost and include built-in capabilities in Microsoft Office, basic GIS tools, base statistics packages, and some advanced tools, such as the NIJ-sponsored CrimeStat series.

Larger agencies with large volumes of incident and intelligence data to be analyzed and shared with many officers will want to consider more sophisticated—and therefore more costly—systems. The key is to think of these as enterprise IT systems making sense of large data sets to provide situational awareness across the department (and, in many cases, to the public) rather than as high-cost crystal balls. The systems should help agencies understand the where, when, and who of crime and identify specific problems driving crime to support law enforcement interventions. These departments will want to ask the following key questions when considering the available tools:

- Is the system easily and effectively able to integrate the department's data from RMS/computer-aided dispatch (CAD) systems and other key sources? The ideal system would allow most incident, suspect, and asset data to be loaded into the system automatically with minimal (if any) manual processing.

- Can the system incorporate not just incident data but also "intelligence" data coming back from the officers (and the community) and provide that information back to officers?
- What is the range of views the system can display, and how can they be tailored for people in different roles? The ideal system would provide "dashboards" tailored to each person's job.
- Does the system display the key data leading to the predictions and not just the predictions? The ideal system will have drilldown capabilities so that staff can see the details of specific crimes, suspicious activity reports, repeat crime locations, and persons of concern leading to high-risk predictions.
- For spatiotemporal predictions, does the system display maps with predicted hot spots clearly shown? Does the system generate hot spots for different times (day/night, weekend/weekday) or in response to different events? Can the system make predictions accounting for where crimes have just occurred—especially for burglaries, which show pronounced and long-lasting near-repeat effects?
- Does the system offer links to data or analysis modules that can help assess personal risk (e.g., the risk that a recently released offender will reoffend, the risk of domestic violence)?
- Does the system support queries across offender data (e.g., social network links, personal appearance, MO) and related data, such as vehicle registries or license plate reads? Similarly, can the system help link related crimes? From a technology perspective, a system with the ability to ingest large amounts of data and support cross-database queries would appear promising for law enforcement purposes.

Later in this chapter, in the "A Crime Fighter's Guide to Predictive Policing: Emerging Practices," we explain how a complete suite of predictive policing tools might work together to create shared situational awareness.

A Developer's Guide to Predictive Policing

Here, we present findings for developers and vendors to help assist their technical and marketing efforts.

Desired Capabilities

The "Buyer's Guide" provided a list of desired capabilities for systems found to be especially useful for predictive policing. We emphasize that systems need to be able to provide situational awareness to different echelons of agencies through tailored "dashboard" displays, alerts, and data drilldown capabilities (along with the necessary database and information-sharing infrastructure needed to support the displays and query responses). The discussion of tools to generate shared situational awareness in the section "A Crime Fighter's Guide to Predictive Policing: Emerging Practices" profiles what a complete suite of interlinked systems might look like.

Looking ahead, moving beyond predictions to provide explicit decision support for resource allocation and other planning decisions could be useful. A recent Police Executive Research Forum survey of 70 member agencies found that while 70 percent have predictive analytics tools, only 22 percent were using the tools to help make resource allocation decisions.[12]

For the most part, the sophistication of the predictive algorithms is secondary. The algorithms need to work reasonably well, but it is not worth, say, multihour running times, to increase measures of performance for predictions by a few percent.

Controlling the Hype

> Dear Sir/Madam: Please let my chief and I know where we can buy the software that will tell us where to go to pick up criminals as they are committing crimes. We have read articles and seen ads on this. . . . (Composite of multiple letters sent to the International Association of Crime Analysts)

We emphasize that predictive policing tools and methods are very useful, but we also emphasize that they are not crystal balls. Unfortunately, communications like the quote above are real, and they are being prompted by media interviews and advertisements that give an impression that one really *can* ask a computer where and when to go to catch criminals in the act. We ask that vendors and developers be accurate in describing their systems as identifying crime risks, not foretelling the future.

Access and Affordability

Developers and vendors must be aware of the major financial limitations that law enforcement agencies face in procuring and maintaining new systems. These limitations are further compounded by major and ongoing budget cuts that are forcing harsh trade-offs among officers and staff, equipment, and technology acquisition.[13] We heard a number of reports from agencies being quoted prices for predictive policing software installation that ran into the high hundreds of thousands or millions of dollars, plus annual maintenance fees of tens or hundreds of thousands of dollars. These costs are simply not affordable for most departments.

That said, we do understand that the investment in developing and maintaining these systems can be quite large, and vendors need to recoup their costs. We strongly

---

[12] Police Executive Research Forum, *How Are Innovations in Technology Transforming Policing?* Washington, D.C., January 2012, p. 1.

[13] In 2012, the Police Executive Research Forum reported that 51 percent of agencies responding to a survey were experiencing budget cuts, and 40 percent expected further cuts in the future. See Police Executive Research Forum, *Policing and the Economic Downturn: Striving for Efficiency Is the New Normal*, Washington, D.C., February 2013.

suggest that vendors consider alternative business models, such as license sharing or partnering with agencies on federal grant applications, to permit smaller agencies greater access to these technologies.

A Crime Fighter's Guide to Using Predictive Policing: Emerging Practices

As mentioned up front, predictive policing is not fundamentally about making crime-related predictions. It is about implementing business processes, such as the policing cycle shown again in Figure 5.4. The objective is to take action to preempt the predicted crime. In this section, we present promising methods to help implement those business processes successfully.

We have noted that there has not been much formal evaluation of predictive policing methods to date, but some efforts (such as NIJ's predictive policing experiments) were under way as of this writing. Nonetheless, from interviewing and viewing presentations from practitioners working on initiatives that might rightly be called predictive policing,[14] and reviewing emerging findings from predictive policing studies, we have identified a number of promising practices for implementing predictive policing business processes. While there is a great deal of variation in the types of predictions being performed, there is also a great deal of agreement on what sorts of measures should be taken to help turn the predictions into prevention.

Figure 5.4
The Prediction-Led Policing Business Process

RAND RR233-5.4

---

[14] We interviewed members of eight law enforcement agencies and qualitatively analyzed the interview notes, coding common themes across the interviews.

Supporting the Establishment of Predictive Policing Processes

Most interviewees stated that top-level support and direction from the relevant chiefs and commanders was necessary for the success of predictive policing initiatives. "Marketing" efforts promoting the value of predictive products, combined with good products that genuinely help solve cases or prevent crimes (and publicizing those successes throughout the department) were also seen as important.

To help expedite the transition to predictive policing, most agencies built their initiatives from existing CompStat models. We repeatedly heard that predictive policing can be seen as the next step from the CompStat concept and the simple reporting of recent crime statistics by operational area.

Typically, predictive policing analysis products are prepared by civilian staff. A number of interviewees stated that resistance to input from civilians by sworn officers was a barrier. The cause of the resistance varied. In some cases, it had to do with prior experiences with improvement initiatives and products that were operationally unsuitable; in others, it had to do with general resistance to using mathematics and computers to help law enforcement or cultural resistance to civilian involvement in law enforcement. Interviewees viewed the combination of top-level support and genuinely beneficial products—preferably for operations that officers were enthused about executing (more on this later)— as the best antidotes to this resistance.

Practices for Data Collection and Analysis: Creating Shared Situational Awareness

We reviewed particular predictive methods and applications earlier in this guide. Here, we focus much more broadly on the concept of using predictions and related information to provide departments with the situational awareness needed to inform operations to interdict crime. Most of our interviewees said their agencies provided officers with some form of situational awareness on recent crimes and criminal intelligence, tailored to specific locations or roles. Tools varied from high-end "business intelligence dashboards" providing tailored, web-based geospatial displays to customized PowerPoint slides or PDF maps disseminated via SharePoint. Officers then typically had the ability to drill down into specific events (e.g., pull complete crime incident reports) or intelligence (e.g., review complete field interview reports) reported in their area.

The information provided might vary, but it generally comes from four key types of tools, ideally interlinked, as shown in Figure 5.5.[15]

---

[15] This concept shown in the figure was originally called the Counterinsurgency Common Operational Picture (COINCOP). COINCOP was developed by several of the authors to show how intelligence could be combined from various sources to create a "common operational picture" in support of counterinsurgency operations. These operations—against insurgents conducting attacks on civilians and coalition forces—share some similarities with policing operations. That said, the architecture shown in Figure 5.5 has been tailored specifically to law enforcement based on our interviews and reviews of the types of systems currently being used. For more information on the original COINCOP concept, see Walter L. Perry and John Gordon IV, *Analytic Support to Intelligence in Counterinsurgencies*, Santa Monica, Calif.: RAND Corporation, MG-682-OSD, 2008, pp. 35–38.

Figure 5.5
Tools to Create Shared Situational Awareness

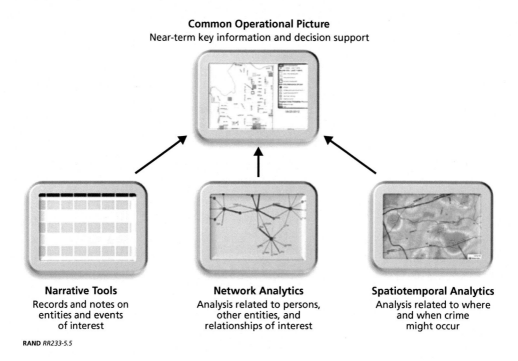

Common Operational Picture
Near-term key information and decision support

Narrative Tools
Records and notes on
entities and events
of interest

Network Analytics
Analysis related to persons,
other entities, and
relationships of interest

Spatiotemporal Analytics
Analysis related to where
and when crime
might occur

RAND *RR233-5.5*

This is meant to be an idealized example, drawn from various systems and displays we have seen to date. As far as we are aware, no system fully matches this concept. The architecture consists of a core awareness tool and three tools providing drilldown information and analytics:

- *A common operational picture (COP) tool:* This is a predominantly geospatial display, tailored for commanders and field officers at all levels to show them what they need to make decisions in near-real time. Data displayed might include
  - recent crimes
  - crime patterns (crimes linked to the same series)
  - specific predictions of near-term interest (projected hot spots/addresses/people in the area)
  - suspicious activity and disorder
  - field interview locations (characterized by contact with a known criminal or someone providing useful intelligence)
  - best-known addresses for outstanding warrants
  - applicable locations for tips and other intelligence
  - information on locations of current and recent police operations (to avoid interference or repetition)
  - current locations of police units

- information on locations of planned police operations
- pop-up boxes providing brief narratives on the above
- icons allowing users to get more detail and analysis from the other tools
- search boxes allowing users to get more detail and analysis from queries.

- *Narrative tools (drilldown capabilities):* These tools bring up detailed records on individuals, locations, vehicles, other assets, organizations, incidents, and events of interest, connecting to RMS and CAD records. Users either click on the corresponding icon in the COP tool or perform a search. New incidents and reports from the field should autopopulate the COP tool. Users should also be able to add additional commentary in response to their findings—ideally, one should think of the records about incidents and entities as similar to wiki entries, updated as needed when new discoveries are made.

- *Network analytics tools:* These tools show (and calculate) relationships between entities and events. Chapter Four discussed how queries across social network data can help identify suspects. Social network analytics can also help commanders and analysts identify key people in a criminal organization or connected to a crime in support of follow-up efforts. Ideally, users should be able to bring up a network of connections corresponding to a particular entity or event from the COP tool; similarly, analysts should be able to export predictions about people, groups, and relationships of interest into the COP.

- *Spatiotemporal predictive analytics tools:* These tools provide specific spatiotemporal analyses, as discussed in Chapters Two and Three, and will be used mostly by analysts. Again, users should be able to bring up the detailed model and forecast by clicking on the appropriate icon in the COP tool, and analysts should be able to export predictions into the COP.

To show an example of what a COP display might look like, Figures 5.6 and 5.7 show sample maps provided to members of the Shreveport Police Department participating in its predictive policing experiment (PILOT). Figure 5.6 shows predicted hot spots (colored grid squares) with little contextual detail. Figure 5.7 adds critical operational context, making it a COP display: It adds overlays created by the Shreveport's crime analysts showing specific locations of interest. Note that Figure 5.7 is a static display, built using ArcGIS, with daily updates. It has no drilldown capabilities, but it also did not require a high-end analytics or business intelligence system. That said, our interviewees agreed that a more sophisticated model, with the ability to provide tailored displays populated with near-real-time data and drilldown capabilities, would be valuable for larger departments.

These are primarily geospatial displays intended primarily for spatiotemporal predictions. Displays for predictions about people are in progress; in addition to maps showing individuals' (and gangs') locations, other displays could include social net-

Figure 5.6
Crime Prediction in District 7, Shreveport, Louisiana

SOURCE: Courtesy of the Shreveport Police Department.
RAND RR233-5.6

Figure 5.7
Crime Prediction in District 7 North, Shreveport, Louisiana, with Contextual Detail

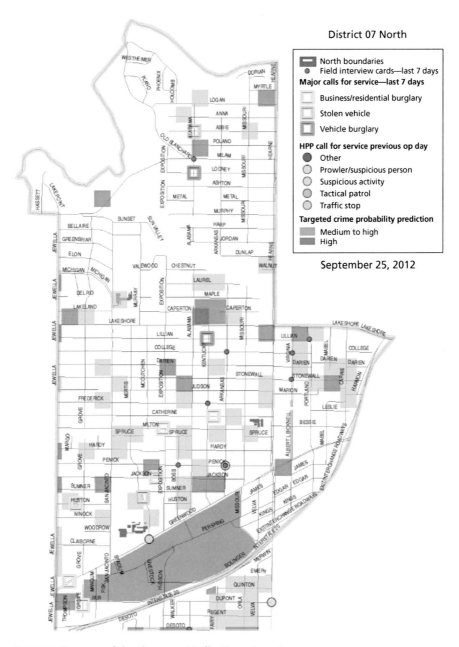

SOURCE: Courtesy of the Shreveport Police Department.
NOTE: HPP = high-priority patrol.

work diagrams, timelines of individual and group activities, key banner alerts, and extended wiki-like reviews of personal histories.

Many of our interviewees stated that a key value of predictive policing systems is in the situational awareness they provide, with statistical predictions being one of many important contributions.

Operations: Key Sources of Information on Interventions

Chapters Two and Three discussed specific interventions and presented case studies of predictive policing in practice. As an addendum to those chapters, this section reviews some key sources of information on different types of interventions. The Office of Justice Programs maintains an online repository of results from evaluations of a wide variety of tested interventions at CrimeSolutions.gov.

For generic interventions (simply allocating more resources to predicted hot spots), we have already discussed the value of the Koper Curve—namely, locating units in a hot spot for 12–16 minutes. For crime-specific interventions, the Center for Problem-Oriented Policing provides numerous guides on measures to help reduce specific types of crimes in a predicted hot spot.

Not surprisingly, problem-specific interventions, employed after specific problem locations and people generating crime risk have been identified, are the most tailored. For locations in which general site characteristics are the problem (as opposed to, say, property owners), the Center for Problem-Oriented Policing provides a guide for making sites safer under the auspices of "crime prevention through environmental design." Examples include installing better lighting, improving sightlines to busy streets, and adding locks or employing other access-control measures. For high-risk individuals, a number of interventions follow "pulling levers" or "focused deterrence" models, which combine multiple types of pressure from police and community groups to change violent behavior, as well as social services to provide support for people who want to change.[16] The National Network for Safe Communities maintains a repository of material on these strategies on its web site. For high-risk personal and gang disputes, Cure Violence (formerly Chicago Ceasefire) uses respected "violence interrupters" to mediate disputes, reduce perceived needs for violence, and connect high-risk individuals with social services.

Promising General Practices for Operations

Here, we focus on general characteristics of successful efforts to take action on predictions. One such characteristic, strong and consistent top-level support, was dis-

---

[16] Pulling levers and focused deterrence policing approaches are built on the idea that relatively small numbers of offenders are responsible for or connected to large numbers of crimes. These strategies target key offenders, discouraging them from committing crimes by making it clear that law enforcement officials are monitoring them.

cussed was discussed earlier in this chapter. The following additional characteristics can improve the likelihood of successful operations:

- *Dedicated resources:* Staff involved in intervention efforts must be given the time and tools (e.g., cars, IT systems) they require. For patrol officers, this typically means freedom to focus on hot spots or "hot people" without having to respond to calls for service for set blocks of time. Similarly, commanders and supporting analysts typically need dedicated time as well.

- *Strong interest and enthusiasm among the staff involved:* In Shreveport commanders were excited about the opportunity to conduct focused patrols in hot spots, saying that it gave them the opportunity to put into action ideas that they had been considering for years. Officers involved in the project were equally enthusiastic about "knocking down crime stats." Similarly, emerging research on hot spot interventions in Philadelphia has found that officers conducting interventions (through focused deterrence/"pulling levers" strategies targeting high-risk persons in hot spots) were enthusiastic about the approach, as it was consistent with what they felt would be most effective.[17]

- *Synchronized support in conducting the operations:* Shreveport's PILOT team could count on support from the lieutenant, field detective, and other officers to immediately help follow up on leads generated from field interviews and other tips. In Baltimore, the police did not just put surveillance cameras in projected hot spots—they have dedicated staff to watch the camera feeds during "hot times" and have dispatchers and officers on call to act immediately when crimes occur on camera.

- *Freedom combined with accountability (but not hostility):* Within a broad intervention framework, most agencies in our study gave commanders and officers substantial flexibility in finding and addressing problems in their hot spots (or with "hot people"). At the same time, they held individuals accountable for preventing crimes and solving crime problems in their area, from the beat to the district level. The focus was not on punishment but on holding regular discussions with officers at all levels, reviewing crime problems they were facing, advising them on fixing crime problems, and ensuring that appropriate actions were taken to address those problems.

- *Working to build good relationships with the community:* Local communities are the biggest source of information on crime and crime threats. A common example of police departments' efforts to build on community relationships was trying to talk with as many people as possible—not to stop and frisk them but to explain that the police were trying to reduce crime in the area. Officers then asked

---

[17] Elizabeth Groff, Temple University, "Police Deployment at Places," presentation at the 2013 Center for Evidence-Based Crime Policy and the Scottish Institute for Policing Research Joint Symposium, Arlington, Va., April 8, 2013.

whether residents could provide any information to support this cause (and provided email and phone contacts to encourage this). Immediately after a burglary, for example, Shreveport police would knock on surrounding neighbors' doors to tell them about the crime and ask whether they had seen anything. Our research also revealed widespread use of focused deterrence measures targeting known criminals acting suspiciously, working with multiple community groups (a core element of "pulling levers" strategies). Further, in measuring progress, departments focused on crime reduction and clearing crime cases, not "raw numbers" (i.e., achieving as many arrests as possible). Interviewees reported seeing more cooperation and collaboration with the public over time. Shreveport personnel stated, for example, that improved relationships with the community were the most important and lasting benefit of their predictive policing effort.

Conclusions

While predictive policing is a new and controversial concept, the application of analytical and quantitative approaches will continue to be an important part of police activities. We have attempted to dispel the most prevalent myths about predictive policing and identify key pitfalls. There are several important takeaways for both practitioners and policymakers.

Practitioners and developers of predictive policing tools must be careful not to overpromise when it comes to the capabilities of their tools and analyses. They should defend the usefulness of these methods, however, and emphasize that they need to be integrated with tactical interventions. Predictive policing begins with data analysis, so it is important that practitioners understand the data and the goal of the analysis. Analysts should gain buy-in from uniformed officers because these officers will be the end users of the predictions. Furthermore, uniformed officers should work closely with analysts to ensure their analyses are tactically useful.

Chiefs and executives should be wary of promises that seem too good to be true. Instead, they should look for approaches that are suitable for their departments. Small departments may not need expensive software, and departments of any size should compare open-source alternatives to commercial products. Larger agencies will want to consider more sophisticated systems. However, the key for agencies of all sizes is to think of the tools as providing situational awareness rather than crystal balls. The systems should help agencies understand the where, when, and who of crime and identify the specific problems driving that criminal activity; this information will help support interventions to address these problems and reduce crime.

Although predictive policing involves advanced mathematical techniques, one need not be a mathematician to understand the basic concepts and implications. Both police executives and policymakers need some understanding of the methodologies so that they can make decisions in ways that support crime reduction and preserve privacy and security.

Finally, we reiterate that predictions are just half of the predictive policing paradigm; taking action is the other half. The following attributes characterize successful interventions:

- There is substantial top-level support for the effort.
- Resources are dedicated to the task.
- The personnel involved are interested and enthusiastic.
- Efforts are made to ensure good working relationships between analysts and officers.
- The predictive policing systems and other department resources provide the shared situational awareness needed to make decisions about where and how to take action.
- Synchronized support is provided when needed.
- Responsible officers have the freedom to carry out interventions, combined with accountability for solving crime problems.
- The interventions are based on building good relationships with the community and good information (intelligence).

Practitioners, especially commanders, need to ensure that appropriate actions are planned and executed in response to predictions and supporting evidence. Policymakers should similarly ask questions about how predictive policing interventions are being planned and performed and which lessons can inform successful strategies in the future.

About the Authors

Walter Perry

Walter Perry (Ph.D., information technology, George Mason University) most recently conducted research identifying the behavioral and social indicators of potential violent acts, focusing primarily on information fusion methods to combine indicator reports. Prior to these studies, he co-developed an algorithm for the Defense Intelligence Agency to indicate when a terrorist group is on the verge of acquiring weapons of mass destruction. He conducted research into methods for developing data fusion and information processing algorithms and analyzing large command-and-control problems through network modeling. He has also developed several metrics of the impact of command and control on military operations for both U.S. Army and U.S. Navy applications. Perry joined RAND in 1984 after a 20-year career with the U.S. Army Signal Corps. He has taught electrical engineering and computer sciences at The George Washington University, statistics at George Mason University, and mathematics at the U.S. Military Academy at West Point.

Brian McInnis

Brian McInnis (M.P.P. education policy, Peabody College, Vanderbilt University) joined RAND in 2009. At RAND, McInnis focuses on K–12 education policy issues and defense acquisition programs and has developed Python-based text mining algorithms for natural language processing. In 2005, he served as a member of the California Preschool Through Post-Secondary Education (P–16) Council, where he focused on issues of K–12 school accountability and academic rigor. While at Vanderbilt, McInnis also developed web-bots and text/data-mining engines in his capacity as a research programmer at the National Center on Performance Incentives.

Carter C. Price

Carter C. Price (Ph.D., applied mathematics, University of Maryland, College Park) is a mathematician at RAND. He has conducted quantitative and qualitative analysis for projects in RAND Health, the RAND National Security Research Division, RAND Arroyo Center, and RAND Justice, Infrastructure, and Environment. His recent work includes an analysis of the Affordable Care Act, a study of the effects of budget cuts on defense, a review of the Transportation Security Administration's risk models, and unmanned aerial system allocation modeling for overwatch. His graduate work focused on increasing the efficiency of health care systems using modeling and simulation.

Susan C. Smith

Susan C. Smith (M.S., management, University of Saint Mary, Kansas; Ph.D. candidate, public policy and administration, specializing in criminal justice) recently retired from law enforcement after a 22-year career, primarily as a crime analyst for the Overland Park and Shawnee, Kansas, police departments. (She contributed to this study while she was at Shawnee.) She is the director of operations at Bair Analytics. She is also an adjunct professor at Johnson County Community College (Overland Park, Kansas), Regis University, Tiffin University, and the University of Missouri, Kansas City, and an adjunct policy analyst at RAND in the NIJ Center of Excellence. She is the president of the International Association of Crime Analysts (IACA) and a past president of the Mid-America Regional Crime Analysis Network. As committee chair of the IACA's annual training conference, she guided the implementation of the IACA Professional Training Series and its Standards, Methods and Technology Committee.

John Hollywood

John Hollywood (Ph.D., operations research, Massachusetts Institute of Technology) is an operations researcher at RAND, a professor of public policy at the Pardee RAND Graduate School (PRGS), and the director of the National Law Enforcement and Corrections Technology Center's Information and Geospatial Technologies Center. His recent research has focused on improving information collection and analysis methods to prevent acts of violence, ranging from violent crime to terrorism and insurgent attacks. Recent projects include evaluating predictive policing experiments for NIJ, examining combat search-and-rescue networks, assessing characteristics of suicide bombing targets in Israel, assessing methods used to foil U.S.-based terrorist plots, and using 911 call data to predict crime hot spots and identify calls potentially related to terrorist surveillance. At PRGS, he teaches a class on applying systems engineering and management practices to public policy.

Bibliography

Adderley, Richard, and Peter B. Musgrove, "Data Mining Case Study: Modeling the Behavior of Offenders Who Commit Serious Sexual Assaults," *Proceedings of the 7th ACM SIGKDD International Conference on Knowledge Discovery and Data Mining*, New York: Association for Computing Machinery, 2001, pp. 215–220.

Albanese, Jay S., "Risk Assessment in Organized Crime: Developing a Market and Product-Based Model to Determine Threat Levels," *Journal of Contemporary Criminal Justice*, Vol. 24, No. 3, August 2008, pp. 263–273.

Allen, John P., and Veronica B. Wilson, eds., "Addiction Severity Index (ASI)," in *Assessing Alcohol Problems: A Guide for Clinicians and Researchers*, 2nd ed., Bethesda, Md.: National Institute on Alcohol and Alcoholism, 2003a. As of August 6, 2013: http://pubs.niaaa.nih.gov/publications/AssessingAlcohol/InstrumentPDFs/04_ASI.pdf

———, "Substance Abuse Subtle Screening Inventory (SASSI)," in *Assessing Alcohol Problems: A Guide for Clinicians and Researchers*, 2nd ed., Bethesda, Md.: National Institute on Alcohol and Alcoholism, 2003b. As of August 6, 2013: http://pubs.niaaa.nih.gov/publications/AssessingAlcohol/InstrumentPDFs/66_SASSI.pdf

Anacapa Sciences, Inc., *Data-Driven Approaches to Crime and Traffic Safety (DDACTS): Case Study of the Metropolitan Nashville, Tennessee, Police Department's DDACTS Program*, Santa Barbara, Calif., October 16, 2009. As of August 6, 2013: http://www.iadlest.org/Portals/0/Files/Documents/DDACTS/Docs/DDACTS_Case_Study-Nashville.pdf

Anselin, Luc, *Local Indicators of Spatial Association (LISA)*, Regional Research Institute Research Paper No. 9331, Morgantown, W. Va.: West Virginia University, 1994.

Apache OpenOffice, homepage, undated. As of August 6, 2013: http://www.openoffice.org

Austin, James, "The Proper and Improper Use of Risk Assessment in Corrections," *Federal Sentencing Reporter*, Vol. 16, No. 3, February 2004, pp. 1–6.

Austin, James, Dana Coleman, Johnette Peyton, and Kelly Dedel Johnson, *Reliability and Validity Study of the LSI-R Risk Assessment Instrument*, Washington, D.C.: Institute on Crimes, Justice and Corrections, George Washington University, January 9, 2003.

Beck, Charlie, and Colleen McCue, "Predictive Policing: What Can We Learn from Wal-Mart and Amazon About Fighting Crime in a Recession?" *The Police Chief*, November 2009.

Bentley, Paul C., "Mapping in Action: Scottsdale Police Department: Starting Up Crime Mapping," *Crime Mapping News*, Vol. 2, No. 1, Winter 2000, pp. 10–11. As of August 6, 2013: http://www.cops.usdoj.gov/html/cd_rom/inaction1/pubs/CrimeMappingNewsletters/Volume2Issue1Winter2000.pdf

Beretich, Tom, "Mapping Programs Target Alcohol-Impaired Driving," *Geography and Public Safety*, Vol. 1, No. 2, July 2008.

Berk, Richard A., *Statistical Learning from a Regression Perspective*, New York: Springer, 2008.

———, "Forecasting Methods in Crime and Justice," *Annual Review of Law and Social Science*, Vol. 4, December 2008, pp. 219–238.

———, "Asymmetric Loss Functions for Forecasting in Criminal Justice Settings," *Journal of Quantitative Criminology*, Vol. 27, No. 1, March 2011, pp. 107–123.

Bess, Michael, "Assessing the Impact of Home Foreclosures in Charlotte Neighborhoods," *Geography and Public Safety*, Vol. 1, No. 3, October 2008.

Boire Filler Group, "Embedding Predictive Analytics into the Corporate Culture," white paper, undated. As of August 6, 2013: http://www.boirefillergroup.com/pdf/Embedding%20Predictive%20Analytics%20into%20the%20Corporate%20Culture.pdf

Bonta, James, and D. A. Andrews, *Risk-Need-Responsivity Model for Offender Assessment and Rehabilitation*, Ottawa, Ont.: Public Safety Canada, 2007.

Borissow, Peter, "Crime Forecast of Washington DC," Wikimedia Commons public domain image, March 30, 2009. As of August 6, 2013: http://en.wikipedia.org/wiki/File:Signature_Analyst_Assessment_of_DC.jpg

Braga, Anthony A., and David L. Weisburd, *Policing Problem Places: Crime Hot Spots and Effective Prevention*, New York: Oxford University Press, 2010.

Bratton, William, John Morgan, and Sean Malinowski, "The Need for Innovation in Policing Today," unpublished manuscript, presented at the Harvard Executive Sessions, October 2009.

Breiman, Leo, "Bagging Predictors," *Machine Learning*, Vol. 24, No. 2, August 1996, pp. 123–140.

———, "Random Forests," *Machine Learning*, Vol. 45, No. 1, October 2001, pp. 5–32.

Brown, Donald, Jason Dalton, and Heidi Hoyle, "Spatial Forecast Methods of Terrorist Events in Urban Environments," *Lecture Notes in Computer Science*, Vol. 3073, 2004, pp. 426–435.

Brown, Donald E., and Steven H. Kerchner, "Spatial-Temporal Point Process Models for Criminal Events," in Donald E. Brown, *Final Report: Predictive Models for Law Enforcement*, prepared for the U.S. Department of Justice, Washington, D.C., November 2002.

Brown v. Plata, U.S. Supreme Court Docket No. 09-1233, 2011.

Bowers, Kate J., Shane D. Johnson, and Ken Pease, "Prospective Hot-Spotting: The Future of Crime Mapping?" *British Journal of Criminology*, Vol. 44, No. 5, 2004, pp. 641–658.

Bushman, Brad J., Morgan C. Wang, and Craig A. Anderson, "Is the Curve Relating Temperature to Aggression Linear or Curvilinear? Assaults and Temperature in Minneapolis Reexamined," *Journal of Personality and Social Psychology*, Vol. 89, No. 1, July 2005, pp. 62–66.

California Department of Justice, Privacy Enforcement and Protection Unit, "Privacy Laws," web page, undated. As of August 6, 2013: http://oag.ca.gov/privacy/privacy-laws

Canter, David, Toby Coffey, Malcolm Huntley, and Chrisopher Missen, "Predicting Serial Killers' Home Base Using a Decision Support System," *Journal of Quantitative Criminology*, Vol. 16, No. 4, December 2000, pp. 457–478.

Canter, David, and Laura Hammond, "A Comparison of the Efficacy of Different Decay Functions in Geographical Profiling for a Sample of US Serial Killers," *Journal of Investigative Psychology and Offender Profiling*, Vol. 3, 2006, pp. 91–103.

Canter, David, and Paul Larkin, "The Environmental Range of Serial Rapists," *Journal of Environmental Psychology*, Vol. 13, No. 1, March 1993, pp. 63–69.

Caplan, Joel M., and Leslie W. Kennedy, eds., *Risk Terrain Modeling Compendium for Crime Analysis*, Rutgers Center on Public Security, Newark, N.J.: Rutgers Center on Public Security, 2011.

Caplan, Joel M., Leslie W. Kennedy, and Joel Miller, "Risk Terrain Modeling: Brokering Criminological Theory and GIS Methods for Crime Forecasting," *Justice Quarterly*, Vol. 28, No. 2, April 2011, pp. 360–381.

Caruana, Rich, and Alexandru Niculescu-Mizil, "An Empirical Comparison of Supervised Learning Algorithms Using Different Performance Metrics," *Proceedings of the 23rd International Conference on Machine Learning*, New York: Association for Computing Machinery, 2006, pp. 161–168.

Cate, Fred H., and Beth E. Cate, "The Supreme Court and Information Privacy," *International Data Policy Law*, Vol. 2, No. 4, 2012, pp. 255–267.

Catherine Hood Consulting, "TRAMO/SEATS FAQ," web page, last updated April 30, 2013. As of August 6, 2013:
http://www.catherinechhood.net/safaqseats.html

Center for Evidence-Based Crime Policy, George Mason University, "Evidence-Based Policing Matrix," web page, undated. As of August 6, 2013:
www.policingmatrix.org

Center for Problem-Oriented Policing, "The SARA Model," web page, undated. As of August 6, 2013:
http://www.popcenter.org/about/?p=sara

Centre for Investigative Psychology, "Dragnet," web page, undated. As of August 6, 2013:
http://www.i-psy.com/publications/publications_dragnet.php

Chainey, Spencer, Lisa Tompson, and Sebastian Uhlig, "The Utility of Hotspot Mapping for Predicting Spatial Patterns of Crime," *Security Journal*, Vol. 21, No. 1, February 2008, pp. 4–28.

Cheung, Jason, and Ryan Prox, Vancouver Police Department, "Fighting Crime with CRIME: A Single, Integrated Analytical, and Investigative Umbrella," presentation to the International Association of Chiefs of Police Law Enforcement Information Management Conference, Indianapolis, Ind., May 23, 2012.

City of Cincinnati, "Neighborhood Enhancement Program," web page, undated. As of August 6, 2013:
http://www.cincinnati-oh.gov/community-development/neighborhood-development/nep

City of Minneapolis, "What Is CODEFOR?" web page, last updated September 27, 2011. As of August 6, 2013:
http://www.minneapolismn.gov/police/statistics/police_about_codefor

Clarke, Ronald V., and Marcus Felson, eds., *Routine Activity and Rational Choice*, New Brunswick, N.J.: Transaction Publishers, 2003.

Cochran, William G., *Sampling Techniques*, 3rd ed., New York: John Wiley and Sons, 1977.

Code of Federal Regulations, Title 28, Judicial Administration, Part 23.30, Operating Principles.

Committee on Technical and Privacy Dimensions of Information for Terrorism Prevention and Other National Goals, National Research Council, *Protecting Individual Privacy in the Struggle Against Terrorists: A Framework for Program Assessment*, Washington, D.C.: National Academies Press, 2008.

Conley, Chris, "Memphis a Victim of Crime Reports," *The Commercial Appeal* (Memphis, Tenn.), June 29, 2009. As of August 6, 2013:
http://www.commercialappeal.com/news/2009/jun/29/memphis-victim-of-crime-reports

Crace, John, "Two Brains," *The Guardian*, November 1, 2004. As of August 6, 2013:
http://www.theguardian.com/uk/2004/nov/02/ukcrime.highereducationprofile

CrimeStat III: A Spatial Statistics Program for the Analysis of Crime Incident Locations, July 2010. As of August 6, 2013:
http://www.icpsr.umich.edu/CrimeStat

Cure Violence, homepage, undated. As of August 6, 2013:
http://cureviolence.org

Dahle, Klaus-Peter, "Strengths and Limitations of Actuarial Prediction of Criminal Reoffence in a German Prison Sample: A Comparative Study of LSI-R, HCR-20 and PCL-R," *International Journal of Law and Psychiatry*, Vol. 29, No. 5, September–October 2006, pp. 431–442.

Davenport, Thomas H., and Jeanne G. Harris, *Competing on Analytics: The New Science of Winning*, Boston, Mass.: Harvard Business School Press, 2007.

Dees, Tim, "Crime Mapping and Geographic Profiling," *LawOfficer*, December 22, 2008. As of August 6, 2013:
http://www.lawofficer.com/article/magazine-feature/crime-mapping-geographic-profi

Eck, John E., Spencer Chainey, James G. Cameron, Michael Leitner, and Ronald E. Wilson, *Mapping Crime: Understanding Hot Spots*, Washington, D.C.: National Institute of Justice, August 2005. As of August 6, 2013:
http://www.nij.gov/topics/technology/maps/ncj209393.htm

Egge, Jeff, "Experimenting with Future-Oriented Analysis at Crime Hot Spots in Minneapolis," *Geography and Public Safety*, Vol. 2, No. 4, March 2011.

Environmental Criminology Research Inc., "Rigel Profiler: Overview," web page, undated. As of August 6, 2013:
http://geographicprofiling.com/products/rigel-profiler

Esri, "ArcGIS for Desktop," web page, undated. As of August 6, 2013:
http://www.esri.com/software/arcgis/arcgis-for-desktop

Estivill-Castro, Vladimir, and Ickjai Lee, "Data Mining Techniques for Autonomous Exploration of Large Volumes of Geo-Referenced Crime Data," in *Proceedings of the 6th International Conference on Geocomputation*, 2001.

Federal Bureau of Investigation, "UCR General FAQs," undated. As of August 6, 2013:
http://www.fbi.gov/about-us/cjis/ucr/frequently-asked-questions/ucr_faqs

———, "Crime in the United States 2005," web page, September 2006. As of August 6, 2013:
http://www.fbi.gov/about-us/cjis/ucr/crime-in-the-u.s/2005

Fenton, Justin, "Tool Gauges Abuse Risk: Program Assesses Danger in Cases of Domestic Violence," *Baltimore Sun*, November 14, 2007. As of August 6, 2013:
http://articles.baltimoresun.com/2007-11-14/
news/0711140029_1_domestic-violence-counselors-lethality

Ferguson, Andrew Guthrie, "Predictive Policing and Reasonable Suspicion," *Emory Law Journal*, Vol. 62, No. 2, 2012, pp. 259–313.

Ferguson, Andrew G., and Damien Bernache, "The 'High-Crime Area' Question: Requiring Verifiable and Quantifiable Evidence for Fourth Amendment Reasonable Suspicion Analysis," *American University Law Review*, Vol. 57, No. 6, August 2008, pp. 1587–1644.

Florida Department of Juvenile Justice, *Fiscal Year 2009–10 Annual Report*, Tallahassee, Fla., 2010.

Friedman, Jerome, Trevor Hastie, and Rob Tibshirani, "glmnet: Lasso and Elastic-Net Regularized Generalized Linear Models," Comprehensive R Archive Network, undated. As of August 6, 2013:
http://cran.r-project.org/web/packages/glmnet

GeoEye Analytics and Alexandria Police Department, *Analysis Report: Elevating Insight for Law Enforcement Using Geospatial Predictive Analytics*, Alexandria, Va., April 1, 2010.

Goode, Erica, "Sending the Police Before There's a Crime," *New York Times*, August 15, 2011. As of August 6, 2013:
http://www.nytimes.com/2011/08/16/us/16police.html

Gorr, Wilpen, and Andreas Olligschlaeger, *Crime Hot Spot Forecasting: Modeling and Comparative Evaluation*, prepared for the U.S. Department of Justice, July 3, 2002.

Gottlieb, Steven, Sheldon Arenberg, and Raj Singh, *Crime Analysis: From First Report to Final Arrest*, Montclair, Calif.: Alpha Group Center for Crime and Intelligence Analysis Training, 1994.

Gotway, C. A., and W. W. Stroup, "A Generalized Linear Model Approach to Spatial Data Analysis and Prediction," *Journal of Agricultural, Biological, and Environmental Statistics*, Vol. 2, No. 2, June 1997, pp. 157–178.

Greenburg, Zack O'Malley, "America's Most Dangerous Cities," *Forbes Magazine*, April 23, 2009. As of August 6, 2013:
http://www.forbes.com/2009/04/23/most-dangerous-cities-lifestyle-real-estate-dangerous-american-cities.html

Greenwald, Mark, "Improving Juvenile Justice for the State of Florida," *Building a Smarter Planet: A Smarter Planet Blog*, April 14, 2010. As of August 6, 2013:
http://asmarterplanet.com/blog/2010/04/improving-juvenile-justice-for-the-state-of-florida.html

Grierson, Bruce, "The Hound of the Data Points," *Popular Science*, March 21, 2003. As of August 6, 2013:
http://www.popsci.com/scitech/article/2003-03/hound-data-points

Groff, Elizabeth, Temple University, "Police Deployment at Places," presentation at the 2013 Center for Evidence-Based Crime Policy and the Scottish Institute for Policing Research Joint Symposium, Arlington, Va., April 8, 2013.

Groff, Elizabeth R., and Nancy G. La Vigne, "Forecasting the Future of Predictive Crime Mapping," in Nick Tilley, ed., *Analysis for Crime Prevention: Crime Prevention Studies*, Vol. 13. Monsey, N.Y.: Criminal Justice Press, 2002, pp. 29–57.

Guay, Jean-Pierre, *Predicting Recidivism with Street Gang Members*, Ottawa, Ont.: Public Safety Canada, 2012.

Gwinn, Samantha L., Christopher Bruce, Julie P. Cooper, and Steven Hick, eds., *Exploring Crime Analysis: Readings on Essential Skills*, 2nd ed., Overland Park, Kan.: International Association of Crime Analysts, 2008.

Haberman, Cory, and Jerry H. Ratcliffe, "The Predictive Policing Challenges of Near Repeat Armed Street Robberies," presentation to the International Crime and Intelligence Analysis Conference, Manchester, UK, November 3–4, 2011.

Hall, Howard B., "Targeting Crash and Crime Hot Spots in Baltimore County," *The Police Chief*, Vol. 76, No. 8, July 2009.

Hall, Howard, and Emily N. Puls, "Implementing DDACTS in Baltimore County: Using Geographic Incident Patterns to Deploy Enforcement," *Geography and Public Safety*, Vol. 2, No. 3, June 2010.

Harrell, Margaret C., and Melissa A. Bradley, *Data Collection Methods: Semi-Structured Interviews and Focus Groups*, Santa Monica, Calif.: RAND Corporation, TR-718-USG, 2009. As of August 6, 2013:
http://www.rand.org/pubs/technical_reports/TR718.html

Harries, Keith, *Mapping Crime: Principle and Practice*, Washington, D.C.: National Institute of Justice, 1999. As of August 6, 2013:
https://www.ncjrs.gov/html/nij/mapping/pdf.html

Harris, Chandler, "Richmond, Virginia, Police Department Helps Lower Crime Rates with Crime Prediction Software," *Government Technology*, December 21, 2008. As of August 6, 2013:
http://www.govtech.com/public-safety/Richmond-Virginia-Police-Department-Helps-Lower.html

Hart, Timothy C., and Paul A. Zandbergen, *Effects of Data Quality on Predictive Hotspot Mapping*, prepared for the National Institute of Justice, Las Vegas, Nev., and Albuquerque, N.M.: University of Nevada, Las Vegas, and University of New Mexico, September 1, 2012.

Hayes, Kerry, "Baltimore Police Department: Incorporating Technology to Reduce Violence," presentation to the International Association of Chiefs of Police Law Enforcement Information Management Conference, Indianapolis, Ind., May 22, 2012.

Heaton, Brian, "Predictive Policing a Success in Santa Cruz, Calif.," *Government Technology*, October 8, 2012. As of August 6, 2013:
http://www.govtech.com/public-safety/Predictive-Policing-a-Success-in-Santa-Cruz-Calif.html

———, "Behavioral Data and the Future of Predictive Policing," *Government Technology*, November 2, 2012. As of August 6, 2013:
http://www.govtech.com/Behavioral-Data-and-the-Future-of-Predictive-Policing.html

Henry, Vincent E., "CompStat Management in the NYPD: Reducing Crime and Improving Quality of Life in New York City," paper presented at the United Nations Asia and Far East Institute for the Prevention of Crime and the Treatment of Offenders 129th International Senior Seminar, 2005. As of August 6, 2013:
http://www.unafei.or.jp/english/pdf/RS_No68/No68_11VE_Henry1.pdf

Hirata, Edson, Osvaldo Almeida, Rossana R. Funari, and Eva L. Klein, "Validity of the Michigan Alcoholism Screening Test (MAST) for the Detection of Alcohol-Related Problems Among Male Geriatric Outpatients," *American Journal of Geriatric Psychiatry*, Vol. 9, No. 1., Winter 2001, pp. 30–34.

Hord, Travis, "Professor Uses Math to Track Criminals, Shark Patterns," *University Star* (Texas State University), August 28, 2009. As of August 6, 2013:
http://star.txstate.edu/content/professor-uses-math-track-criminals-shark-patterns

Horney, Julie, D. Wayne Osgood, and Ineke Haen Marshall, "Criminal Careers in the Short-Term: Intra-Individual Variability in Crime and Its Relation to Local Life Circumstances," *American Sociological Review*, Vol. 60, No. 5, October 1995, pp. 655–673.

"HotSpot Detective v.2," homepage, undated. As of August 6, 2013:
http://jratcliffe.net/hsd/index.htm

IBM, "SPSS software," web page, undated. As of August 6, 2013:
http://www-01.ibm.com/software/analytics/spss

———, "Predictive Analytics—Police Use Analytics to Reduce Crime," video advertisement, March 27, 2012. As of August 6, 2013:
http://www.youtube.com/watch?v=_ZyU6po_E74

International Association of Crime Analysts, *Crime Pattern Definitions for Tactical Analysis*, White Paper 2011-01, Overland Park, Kan., 2011.

International Association of Directors of Law Enforcement Standards and Training, homepage, undated. As of August 6, 2013:
https://www.iadlest.org

Institute for Intergovernmental Research, "Nationwide SAR Initiative: Resources," web page, undated. As of August 6, 2013:
http://nsi.ncirc.gov/resources.aspx

"The Interrupters," television broadcast, *PBS Frontline*, February 14, 2012. As of August 6, 2013:
http://www.pbs.org/wgbh/pages/frontline/interrupters

Johnson, Shane D., Kate J. Bowers, Dan J. Birks, and Ken Pease, "Predictive Mapping of Crime by ProMap: Accuracy, Units of Analysis, and the Environmental Backcloth," in David Weisburd, Wim Bernasco, and Gerben J. N. Bruinsma, eds., *Putting Crime in Its Place: Units of Analysis in Geographic Criminology*, New York: Springer, 2009, pp. 171–198.

Kanable, Rebecca, "Dig into Data Mining," *Officer.com*, April 12, 2007. As of August 6, 2013:
http://www.officer.com/article/10250000/dig-into-data-mining

Katrandjian, Olivia, "Hurricane Irene: Pop-Tarts Top List of Hurricane Purchases," ABC News, August 27, 2011. As of August 6, 2013:
http://abcnews.go.com/US/hurricanes/hurricane-irene-pop-tarts-top-list-hurricane-purchases/story?id=14393602

Kennedy, David, "After a Horrific Summer of Murder, Chicago Trying a Bold New Approach," *The Daily Beast*, September 28, 2012. As of August 6, 2013:
http://www.thedailybeast.com/articles/2012/09/28/after-a-horrific-summer-of-murder-chicago-trying-a-bold-new-approach.html

Koehn, Josh, "Algorithmic Crime Fighting," *SanJose.com*, February 22, 2012. As of August 6, 2013:
http://www.sanjose.com/news/2012/02/22/sheriffs_office_fights_property_crimes_with_predictive_policing

Koper, Christopher S., "Just Enough Police Presence: Reducing Crime and Disorderly Behavior by Optimizing Patrol Time in Crime Hot Spots," *Justice Quarterly*, Vol. 12, No. 4, 1995, pp. 649–672.

Koppelman, Tess, "Lethality Assessment Leads Women to Domestic Violence Shelter," WDAF-TV, Fox 4 News, Kansas City, July 23, 2012. As of August 6, 2013:
http://fox4kc.com/2012/07/23/lethality-assessment-leads-women-to-domestic-violence-shelter

Kreyzig, Erwin, *Introductory Functional Analysis*, Hoboken, N.J.: John Wiley and Sons, 1978.

Levine, Ned, "Crime Mapping and the Crimestat Program," *Geographical Analysis*, Vol. 38, No. 1, January 2006, pp. 41–56.

———, "The 'Hottest' Part of a Hotspot: Comments on 'The Utility of Hotspot Mapping for Predicting Spatial Patterns of Crime,'" *Security Journal*, Vol. 21, No. 4, October 2008, pp. 295–302.

Liu, Hua, and Donald E. Brown, "A New Point Process Transition Density Model for Space-Time Event Prediction," in Donald E. Brown, *Final Report: Predictive Models for Law Enforcement*, prepared for the U.S. Department of Justice, Washington, D.C., 2002.

Liptak, Adam, "Justices, 5-4, Tell California to Cut Prison Population," *New York Times*, May 23, 2011. As of August 6, 2013:
http://www.nytimes.com/2011/05/24/us/24scotus.html

Loeber, Rolf, Dustin Pardini, D. Lynn Homish, Evelyn H. Wei, Anne M. Crawford, David P. Farrington, Magda Stouthamer-Loeber, Judith Creemers, Steven A. Koehler, and Richard Rosenfeld, "The Prediction of Violence and Homicide in Young Men," *Journal of Consulting and Clinical Psychology*, Vol. 73, No. 6, 2005, pp. 1074–1088.

Lum, Cynthia, Christopher S. Koper, and Cody W. Telep, "The Evidence-Based Policing Matrix," *Journal of Experimental Criminology*, Vol. 7, No. 1, March 2011, pp. 3–26.

Machine Learning Group, University of Waikato, "Weka 3: Data Mining Software in Java," homepage, undated. As of August 6, 2013:
http://www.cs.waikato.ac.nz/ml/weka

Marshall, Mark A., "N-DEx: The National Information Sharing Imperative," *The Police Chief*, Vol. 73, No. 6, 2006.

Maryland Network Against Domestic Violence, "What Is LAP?" web page, undated. As of August 6, 2013:
http://mnadv.org/lethality/what-is-lap

McCue, Colleen, *Data Mining and Predictive Analytics: Intelligence Gathering and Crime Analysis*, Burlington, Mass.: Butterworth-Heinemann, 2006.

McCue, Colleen, Lehew Miller, and Steve Lambert, "The Northern Virginia Military Shooting Series: Operational Validation of Geospatial Predictive Analytics," *The Police Chief*, Vol. 80, No. 2, February 2013.

McCue, Colleen, and Andre Parker, "Connecting the Dots: Data Mining and Predictive Analytics in Law Enforcement and Intelligence Analysis," *The Police Chief*, Vol. 10, No. 10, October 2003. As of August 6, 2013:
http://www.policechiefmagazine.org/magazine/index.cfm?fuseaction=display_arch&article_id=121&issue_id=102003

McKinney, Matt, "Targeting the Next Crime," *Minneapolis Star Tribune*, January 26, 2011. As of August 6, 2013:
http://www.startribune.com/local/minneapolis/114425994.html

———, "Taking Back Peavey Park," *Minneapolis Star Tribune*, July 27, 2011. As of August 6, 2013:
http://www.startribune.com/local/minneapolis/126151933.html

Merriam-Webster, Inc., "Data Mining," online dictionary entry, undated. As of August 6, 2013:
http://www.merriam-webster.com/dictionary/data%20mining

Microsoft Corporation, "Excel 2010," homepage, undated. As of August 6, 2013:
http://office.microsoft.com/en-us/excel

Mohler, G. O., M. B. Short, P. J. Brantingham, F. P. Schoenberg, and G. E. Tita, "Self-Exciting Point Process Modeling of Crime," *Journal of the American Statistical Association*, Vol. 106, No. 493, 2011, pp. 100–108.

Monahan, John, *Predicting Violent Behavior: An Assessment of Clinical Techniques*, Thousand Oaks, Calif.: Sage Publications, 1981.

Moore, Andrew, "Statistical Data Mining Tutorials," Auton Lab, Carnegie Mellon University, 2006. As of August 6, 2013:
http://www.autonlab.org/tutorials

Morselli, Carlo, Mathilde Turcotte, and Valentian Tenti, *The Mobility of Criminal Groups*, Ottawa, Ont.: Public Safety Canada, 2010. As of August 6, 2013:
http://publications.gc.ca/collections/collection_2012/sp-ps/PS4-91-2010-eng.pdf

Muller, John K., *Preview of Predictive Policing*, IBM Spatial Solutions, September 13, 2011. As of August 6, 2013:
https://www.ibm.com/developerworks/community/blogs/johnmuller/entry/preview_of_predictive_policing_publication11?lang=en

Multi-Health Systems, Inc., "LS/CMI: Product Overview," web page, undated. As of August 6, 2013:
http://www.mhs.com/product.aspx?gr=saf&prod=ls-cmi&id=overview

National Center for Injury Prevention and Control, Centers for Disease Control and Prevention, "Youth Violence Datasheet," Atlanta, Ga., 2012. As of August 6, 2013:
http://www.cdc.gov/violenceprevention/pdf/yv-datasheet-a.pdf

National Criminal Justice Association, *Justice Information Privacy Guideline: Developing, Drafting and Assessing Privacy Policy for Justice Information Systems*, Washington, D.C., September 2002.

National Institute of Justice, "Geographic Profiling," web page, last updated December 15, 2009. As of August 6, 2013:
http://www.nij.gov/topics/technology/maps/gp.htm

———, "Predictive Policing Symposium: The Future of Prediction in Criminal Justice," web page, last updated December 18, 2009. As of August 6, 2013:
http://www.nij.gov/topics/law-enforcement/strategies/predictive-policing/symposium/future.htm

———, "Predictive Policing Symposiums: What Chiefs Expect from Predictive Policing—Perspectives from Police Chiefs," web page, last updated January 6, 2012. As of August 6, 2013:
http://www.nij.gov/nij/topics/law-enforcement/strategies/predictive-policing/symposium/what-we-expect.htm

National Network for Safe Communities, homepage, undated. As of August 6, 2013:
http://www.nnscommunities.org

Ned Levine and Associates, *CrimeStat III: A Spatial Statistics Program for the Analysis of Crime Incident Locations*, Washington, D.C.: National Institute of Justice, November 2004.

Neill, Daniel B., and Wilpen L. Gorr, "Detecting and Preventing Emerging Epidemics of Crime," *Advances in Disease Surveillance*, Vol. 4, No. 13, 2007.

"New Police Motto: To Predict and Serve?" video, *NBC Nightly News*, August 20, 2011. As of August 6, 2013:
http://www.nbcnews.com/id/21134540/vp/44214509

Ngai, E. W. T., Yong Hu, Y. H. Wong, Yijun Chen, and Xin Sun, "The Application of Data Mining Techniques in Financial Fraud Detection: A Classification Framework and an Academic Review of Literature," *Decision Support Systems*, Vol. 50, No. 3, February 2011, pp. 559–569.

Nucleus Research, *ROI Case Study: IBM SPSS—Memphis Police Department*, Boston, Mass., Document K31, June 2010.

Nyce, Charles, *Predictive Analytics White Paper*, Malvern, Pa.: American Institute for Chartered Property Casualty Underwriters/Insurance Institute of America, 2007. As of August 6, 2013: http://www.theinstitutes.org/doc/predictivemodelingwhitepaper.pdf

Oberwittler, Dietrich, and Marc Wiesenhütter, "The Risk of Violent Incidents Relative to Population Density in Cologne Using the Dual Kernel Density Routine," in Ned Levine and Associates, *CrimeStat III: A Spatial Statistics Program for the Analysis of Crime Incident Locations*, prepared for the National Institute of Justice, November 2004.

Okabe, Atsuyuki, Barry Boots, Kokichi Sugihara, and Song Nok Chiu, *Spatial Tessellations: Concepts and Applications of Voronoi Diagrams*, 2nd ed., New York: John Wiley and Sons, 2000.

Ord, J. K., and Arthur Getis, "Local Spatial Autocorrelation Statistics: Distributional Issues and an Application," *Geographical Analysis*, Vol. 27, No. 4, October 2005, pp. 286–306.

Organisation for Economic Co-operation and Development, "OECD Guidelines on the Protection of Privacy and Transborder Flows of Personal Data: Background," September 23, 1980. As of August 6, 2013: http://www.oecd.org/internet/ieconomy/oecdguidelinesontheprotectionofprivacyandtransborderflowsofpersonaldata.htm

Ouelette, Danielle, "A Hot Spots Experiment: Sacramento Police Department," *Community Policing Dispatch*, Vol. 5, No. 6, June 2012. As of August 6, 2013: http://cops.usdoj.gov/html/dispatch/06-2012/hot-spots-and-sacramento-pd.asp

Papachristos, Andrew V., Anthony A. Braga, and David M. Hureau, "Social Networks and the Risk of Gunshot Injury," *Journal of Urban Health*, Vol. 89, No. 6, December 2012, pp. 992–1003.

Parker, Nora, "Portland Police Bureau Makes Geospatial Widely Accessible," *Directions Magazine*, February 11, 2008. As of August 6, 2013: http://www.directionsmag.com/articles/portland-police-bureau-makes-geospatial-widely-accessible/122762

Pearsall, Beth, "Predictive Policing: The Future of Law Enforcement?" *National Institute of Justice Journal*, No. 266, May 2010.

Pearson Education, "Beck Depression Inventory®–II (BDI®-II)," web page, undated. As of August 6, 2013: http://www.pearsonassessments.com/HAIWEB/Cultures/en-us/Productdetail.htm?Pid=015-8018-370

———, "Hare Psychopathy Checklist–Revised (PCL-R), 2nd Edition," web page, undated. As of August 6, 2013: http://www.pearsonassessments.com/HAIWEB/Cultures/en-us/Productdetail.htm?Pid=PAapclr&Mode=summary

———, "Minnesota Multiphasic Personality Inventory®–2 (MMPI®-2)," web page, undated. As of August 6, 2013: http://psychcorp.pearsonassessments.com/HAIWEB/Cultures/en-us/Productdetail.htm?Pid=MMPI-2&Mode=summary

Pell, Stephanie K., "Systematic Government Access to Private-Sector Data in the United States," *International Data Privacy Law*, Vol. 2, No. 4, 2012.

Perry, Walter L., and John Gordon IV, *Analytic Support to Intelligence in Counterinsurgencies*, Santa Monica, Calif.: RAND Corporation, MG-682-OSD, 2008. As of August 6, 2013:
http://www.rand.org/pubs/monographs/MG682.html

Peterson, Marilyn, *Intelligence-Led Policing: The New Intelligence Architecture*, Washington, D.C.: U.S. Department of Justice, Bureau of Justice Assistance, Report No. 210681, September 2005. As of August 6, 2013:
https://www.ncjrs.gov/pdffiles1/bja/210681.pdf

Pew Center on the States, *Prison Count 2010: State Population Declines for the First Time in 38 Years*, Washington, D.C., April 1, 2010. As of August 6, 2013:
http://www.pewstates.org/research/reports/prison-count-2010-85899372907

Phelps, Candy, "IBM Predictive Analytics Help Slash Crime Rates in Memphis," *Public Safety IT Magazine*, November 2010. As of August 6, 2013:
http://www.hendonpub.com/resources/article_archive/results/details?id=1530

Police Executive Research Forum, *How Are Innovations in Technology Transforming Policing?* Washington, D.C, January 2012. As of August 6, 2013:
http://policeforum.org/library/critical-issues-in-policing-series/Technology_web2.pdf

————, *Policing and the Economic Downturn: Striving for Efficiency Is the New Normal*, Washington, D.C., February 2013. As of August 6, 2013:
http://policeforum.org/library/critical-issues-in-policing-series/Economic_Downturn.pdf

Predictive Analytics World, "Predictive Analytics Resource Guide," web page, undated. As of August 6, 2013:
http://www.predictiveanalyticsworld.com/predictive_analytics.php

"'Predictive Policing' Making LA Safer," video, *CBS Evening News*, April 11, 2012. As of August 6, 2013:
http://www.cbsnews.com/video/watch/?id=7404996n

Public Law 90-351, Omnibus Crime Control and Safe Streets Act, June 19, 1968.

The R Project for Statistical Computing, homepage, undated. As of August 6, 2013:
http://www.r-project.org

Raïffa, Howard, *Decision Analysis: Introductory Lectures on Choices Under Uncertainty*, Boston, Mass.: Addison-Wesley, 1968.

Ratcliffe, Jerry H., "Near Repeat Calculator," version 1.3, Philadelphia, Pa., and Washington, D.C.: Temple University and National Institute of Justice, August 2009. As of August 6, 2013:
http://www.temple.edu/cj/misc/nr/download.htm

Remsberg, Charles, "Lethality Assessment Helps Gauge Danger from Domestic Disputes," *PoliceOne.com*, December 12, 2007. As of August 6, 2013:
http://www.policeone.com/patrol-issues/articles/1638964-Lethality-Assessment-helps-gauge-danger-from-domestic-disputes

Rich, Thomas, "Crime Mapping and Analysis by Community Organizations in Hartford, Connecticut," *National Institute of Justice Research in Brief*, March 2001. As of August 6, 2013:
https://www.ncjrs.gov/pdffiles1/nij/185333.pdf

Rich, Tom, and Michael Shively, *A Methodology for Evaluating Geographic Profiling Software*, draft report, Cambridge, Mass.: Abt Associates, prepared for the U.S. Department of Justice, December 2004.

Ridgeway, Greg, "The Pitfalls of Prediction," *National Institute of Justice Journal*, No. 271, February 2013. As of August 6, 2013:
http://www.nij.gov/nij/journals/271/pitfalls-of-prediction.htm

Rose, Kristina, Acting Director, National Institute of Justice, "Predictive Policing Symposium: Opening Remarks," transcript of speech at Predictive Policing Symposium, Los Angeles, Calif., November 19, 2009. As of August 6, 2013:
http://www.nij.gov/nij/topics/law-enforcement/strategies/predictive-policing/symposium/opening-rose.htm

Rossmo, D. Kim, *Geographic Profiling: Target Patterns of Serial Murders*, doctoral thesis, Burnaby, B.C.: Simon Fraser University, 1995. As of August 6, 2013:
http://ir.lib.sfu.ca/handle/1892/8121

Rubin, Joel, "Stopping Crime Before It Starts," *Los Angeles Times*, August 21, 2010. As of August 6, 2013:
http://articles.latimes.com/2010/aug/21/local/la-me-predictcrime-20100427-1

SAS, "SAS 9.4," web page, undated. As of August 6, 2013:
http://www.sas.com/software/sas9

Schapire, Robert E., "The Boosting Approach to Machine Learning: An Overview," in David D. Denison, Mark H. Hansen, Christopher C. Holmes, Bani Mallic, and Bin Yu, eds., *Lecture Notes in Statistics: Nonlinear Estimation and Classification*, Vol. 171, New York: Springer, 2003, pp. 149–172.

Schneider, Stephen, *Predicting Crime: The Review of Research*, Ottawa, Ont.: Canadian Department of Justice, 2002. As of August 6, 2013:
http://www.justice.gc.ca/eng/rp-pr/csj-sjc/jsp-sjp/rr02_7/index.html

Sentinel Software Group, "Introduction to the Sentinel Pattern Identification and Correlation Engine," unpublished white paper, 2009.

Shafer, Glenn, *A Mathematical Theory of Evidence*, Princeton, N.J.: Princeton University Press, 1976.

Shannon, Claude, "A Mathematical Theory of Communication," *Bell System Technical Journal*, Vol. 27, July and October 1948, pp. 379–423 and 623–656.

Shelby County District Attorney General, "Four Area Hotels Closed for Business Following 'Operation Heartbreak Hotel,'" news release, February 12, 2008.

Silverii, Scott, "Changing the Culture and History of Policing," *Law Enforcement Today*, October 9, 2012. As of August 6, 2013:
http://lawenforcementtoday.com/2012/10/09/changing-the-culture-and-history-of-policing

Smarandache, Florentin, and Jean Dezert, eds., *Advances and Applications of DSmT for Information Fusion (Collected Works)*, Rehoboth, N.M.: American Research Press, 2009.

Smith, Jeffery, "Memphis Police Leverage Analytics to Fight Crime," *CivSource*, July 21, 2010. As of August 6, 2013:
http://civsourceonline.com/2010/07/21/memphis-police-leverage-analytics-to-fight-crime

Smith, Susan C., and Christopher W. Bruce, *CrimeStat III User Workbook*, Washington, D.C., National Institute of Justice, June 2008. As of August 6, 2013:
http://www.icpsr.umich.edu/CrimeStat/workbook.html

Snook, Brent, Paul J. Taylor, and Craig Bennell, "Geographic Profiling: The Fast, Frugal, and Accurate Way," *Applied Cognitive Psychology*, Vol. 18, No. 1, January 2004, pp. 105–121.

Snook, Brent, Michele Zito, Craig Bennell, and Paul J. Taylor, "On the Complexity and Accuracy of Geographic Profiling Strategies," *Journal of Quantitative Criminology*, Vol. 21, No. 1, March 2005, pp. 1–26.

Stanton, Jeffrey M., "Galton, Pearson, and the Peas: A Brief History of Linear Regression for Statistics Instructors," *Journal of Statistics Education*, Vol. 9, No. 3, 2001.

Stark, Rodney, "Deviant Places: A Theory of the Ecology of Crime," *Criminology*, Vol. 25, No. 4, November 1987, pp. 893–909.

Sullivan, T. J., and Walter L. Perry, "Identifying Indicators of Chemical, Biological, Radiological, and Nuclear (CBRN) Weapons Development Activity in Sub-National Terrorist Groups," *Journal of the Operational Research Society*, Vol. 55, No. 4, April 2004, pp. 361–374.

Sun, Key, *Correctional Counseling: A Cognitive Growth Perspective*, 2nd ed., Burlington, Mass.: Jones and Bartlett Learning, 2003.

Tayebi, Mohammad A., Uwe Glässer, and Patricia L. Brantingham, *Organized Crime Detection in Co-offending Networks*, working paper, Burnaby, B.C.: Simon Fraser University, c. 2011. As of August 6, 2013:
http://www.erdr.org/textes/tayebi_glasser_brantingham.pdf

Telep, Cody W., "Police Interventions to Reduce Violent Crime: A Review of Rigorous Research," Fairfax, Va.: Center for Evidence-Based Crime Policy, George Mason University, 2009. As of August 6, 2013:
http://gemini.gmu.edu/cebcp/Briefings/Telep.pdf

Thompson, Kalee, "The Santa Cruz Experiment: Can a City's Crime Be Predicted?" *Popular Science*, November 1, 2011. As of August 6, 2013:
http://www.popsci.com/science/article/2011-10/santa-cruz-experiment

Townsley, Michael, Ross Homel, and Janet Chaseling, "Repeat Burglary Victimisation: Spatial and Temporal Patterns," *Australian and New Zealand Journal of Criminology*, Vol. 33, No. 1, April 2000, pp. 37–63.

"Transcript: Perspectives in Law Enforcement—The Concept of Predictive Policing: An Interview with Chief William Bratton," U.S. Department of Justice, Bureau of Justice Assistance, November 2009.

Transtutors, "Difference Between Forecasting and Prediction," undated. As of August 6, 2013:
http://www.transtutors.com/homework-help/industrial-management/forecasting/difference-between-forecasting-and-prediction.aspx

Tusikov, Natasha, "Measuring Organised Crime–Related Harms: Exploring Five Policing Methods," *Crime, Law, and Social Change*, Vol. 57, No. 1, February 2012, pp. 99–115.

Uchida, Craig D., *A National Discussion of Predictive Policing: Defining Our Terms and Mapping Successful Implementation Strategies*, prepared for the National Institute of Justice, Document No. NCJ 230404, May 2010. As of August 6, 2013:
https://www.ncjrs.gov/pdffiles1/nij/grants/230404.pdf

United Nations Office on Drugs and Crime, *Criminal Intelligence: Manual for Analysts*, Vienna, Austria, April 2011. As of August 6, 2013:
http://www.unodc.org/documents/organized-crime/Law-Enforcement/Criminal_Intelligence_for_Analysts.pdf

United States v. Jones, U.S. Supreme Court Docket No. 10-1259, 2012.

United States v. Miller, 425 U.S. 435, 1976.

U.S. Census Bureau, "The X-12-ARIMA Seasonal Adjustment Program," web page, undated. As of August 6, 2013:
http://www.census.gov/srd/www/x12a

U.S. Department of Justice, *The National Criminal Intelligence Sharing Plan*, Washington, D.C., October 2003.

U.S. Department of Justice, Office of Justice Programs, "Crime and Crime Prevention: Violent Crime," web page, undated. As of August 6, 2013:
http://www.crimesolutions.gov/TopicDetails.aspx?ID=25

Valentino-DeVries, Jennifer, "How Technology Is Testing the Fourth Amendment," *Wall Street Journal*, September 21, 2011. As of August 6, 2013:
http://blogs.wsj.com/digits/2011/09/21/how-technology-is-testing-the-fourth-amendment

Vander Beken, Tom, "Risky Business: A Risk-Based Methodology to Measure Organized Crime," *Crime, Law, and Social Change*, Vol. 44, No. 5, June 2004, pp. 471–516.

Wang, Xiaofeng, and Donald E. Brown, "The Spatio-Temporal Modeling for Criminal Incidents," *Security Informatics*, Vol. 1, No. 2, February 2012.

Wartell, Julie, Independent Adviser on Public Safety, "GIS for Proactive Policing and Crime Analysis," presentation at the Technologies for Critical Infrastructure Protection Conference, National Harbor, Md., August 31, 2011.

Wartell, Julie, and J. Thomas McEwen, *Privacy in the Information Age: A Guide for Sharing Crime Maps and Spatial Data*, Washington, D.C.: National Institute of Justice, July 2001.

Wascalus, Jacob, Jeff Matson, and Michael Grover, *Assembly and Uses of a Data-Sharing Network in Minneapolis*, Washington, D.C.: Board of Governors of the Federal Reserve System, last updated April 4, 2012. As of August 6, 2013:
http://www.federalreserve.gov/publications/putting-data-to-work-data-sharing-network.htm

Weisburd, David, Gerben J. N. Bruinsma, and Wim Bernasco, "Units of Analysis in Geographic Criminology: Historical Development, Critical Issues, and Open Questions," in David Weisburd, Wim Bernasco, Gerben J. N. Bruinsma, eds., *Putting Crime in Its Place: Units of Analysis in Geographic Criminology*, New York: Springer, 2009, pp. 3–34.

"What Is Predictive Policing?" *WP's Police Tech*, March 16, 2012. As of August 6, 2013:
http://tech.wiredpig.us/post/12291823038/what-is-predictive-policing

Williams, Amy O., "Blue C.R.U.S.H. Walks Its Beat Among Community Organizations," *Daily News* (Memphis, Tenn.), November 16, 2006. As of August 6, 2013:
http://www.memphisdailynews.com/editorial/Article.aspx?id=31455

Wilson, Ronald E., "Place as the Focal Point: Developing a Theory for the DDACTS Model," *Geography and Public Safety*, Vol. 2, No. 3, June 2010.

Wilson, Ronald E., and Derek J. Paulsen, "Foreclosures and Crime: A Geographical Perspective," *Geography and Public Safety*, Vol. 1, No. 3, October 2008.

Wilson, Ronald E., Susan C. Smith, John D. Markovic, and James L. LeBeau, *Geospatial Technology Working Group Meeting Report on Predictive Policing*, Scottsdale, Ariz., October 2009. As of August 6, 2013:
http://www.nij.gov/nij/topics/law-enforcement/strategies/predictive-policing/geospatial-twg-predictive-policing-meeting-summary.pdf

Witten, Ian H., and Eibe Frank, *Data Mining: Practical Machine Learning Tools and Techniques*, 2nd ed., Burlington, Mass.: Morgan Kaufmann, 2005.

Wyatt, Jason, "Integrating Crime and Traffic Crash Data in Nashville," *Geography and Public Safety*, Vol. 2, No. 3, June 2010.

Yale University, Department of Statistics, "Linear Regression," supplemental course materials, September 16, 1997. As of August 6, 2013:
http://www.stat.yale.edu/Courses/1997-98/101/linreg.htm

Zetter, Kim, "U.S. Cities Relying on Precog Software to Predict Murder," *Wired*, January 10, 2013. As of August 6, 2013:
http://www.wired.com/threatlevel/2013/01/precog-software-predicts-crime

Zou, Hui, and Trevor Hastie, "Regularization and Variable Selection via the Elastic Net," *Journal of the Royal Statistics Society: Series B (Statistical Methodology)*, Vol. 67, No. 2, April 2005, pp. 301–320.

Zoutendijk, Andries Johannes, "Organized Crime Threat Assessments: A Critical Review," *Crime, Law, and Social Change*, No. 54, No. 1, August 2010, pp. 63–86.

Zucchini, Walter, *Applied Smoothing Techniques, Part 1: Kernel Density Estimation*, Philadelphia, Pa.: Temple University, October 2003. As of August 6, 2013:
http://isc.temple.edu/economics/Econ616/Kernel/ast_part1.pdf